PIC in Practice

PIC in Practice

D. W. Smith

Newnes

OXFORD AMSTERDAM BOSTON LONDON NEW YORK PARIS
SAN DIEGO SAN FRANCISCO SINGAPORE SYDNEY TOKYO

Newnes
An imprint of Elsevier Science
Linacre House, Jordan Hill, Oxford OX2 8DP
225 Wildwood Avenue, Woburn, MA 01801-2041

First published 2002

British Library Cataloguing in Publication Data
A catalogue record for this book is available from the British Library

ISBN 0 7506 4812 0

For information on all Newnes publications visit
our website at newnespress.com

Typeset by Avocet Typeset, Brill, Aylesbury, Bucks
Printed in Malta by Gutenberg Press Ltd

Contents

Introduction

The microcontroller is an exciting new device in the field of electronics control. It is a complete computer control system on a single chip. Microcontrollers include EPROM program memory, user RAM for storing program data, timer circuits, an instruction set, special function registers, power on reset, interrupts, low power consumption and a security bit for software protection. Some microcontrollers such as the 16C7X devices include on-board A to D converters.

The microcontroller is used as a single chip control unit, for example in a washing machine, and the inputs to the controller would be from a door catch, water level switch, or temperature sensor. The outputs would then be fed to a water inlet valve, heater, motor and pump. The controller would monitor the inputs and decide which outputs to switch on, i.e. close the door – water inlet valve open – monitor water level, close valve when water level reached, check temperature, turn on heater, switch off heater when the correct temperature is reached, turn the motor slowly clockwise for 5 seconds, anticlockwise for 5 seconds, repeat 20 times, etc. Microcontrollers can be used to control other devices, for example disco lights.

The microcontroller because of its versatility, ease of use and cost will change the way electronic circuits are designed and will now enable projects to be designed which previously were too complex. Additional components such as Versatile Interface Adapters (VIA), RAM, ROM, EPROM and address decoders are no longer required.

One of the most difficult hurdles to overcome when using any new technology is the first one – getting started! It was my aim when writing this book to explain as simply as possible how to program and use the PIC microcontrollers. I hope I have succeeded.

Dave Smith, BSc, MSc
Senior Lecturer in Electronics,
Manchester Metropolitan University

1
Introduction to the PIC microcontroller

A microcontroller is a computer control system on a single chip. It has many electronic circuits built into it, which can decode written instructions and convert them to electrical signals. The microcontroller will then step through these instructions and execute them one by one. As an example a microcontroller could be instructed to measure the temperature of a room and turn on a heater if it goes cold. Microcontrollers are now changing electronics design. Instead of hard wiring a number of logic gates together to perform some function we now use instructions to wire the gates electronically. The list of these instructions given to the microcontroller is called a program.

The aim of the book

The aim of the book is to teach you how to build control circuits using devices such as switches, keypads, analogue sensors, LEDs, buzzers, 7-segment displays, alphanumeric displays, radio transmitters, etc. This is done by introducing graded examples starting with only a few instructions and gradually increasing the number of instructions as the complexity of the examples increases.

Each chapter clearly identifies the new instructions added to your vocabulary. The programs use building blocks of code that can be reused in many different program applications. Complete programs are provided so that an application can be seen working. The reader is then encouraged to modify the code to alter the program in order to enhance their understanding.

Throughout this book the programs are written in a language called assembly language which uses a vocabulary of 35 words called an instruction set. In order to write a program we need to understand what these words mean and how we can combine them. The complete instruction set is shown in Chapter 16 Instruction set, files and registers.

All of the programs illustrated in the book are available from the Butterworth-Heinemann website on www.bh.com. You will of course need a programmer to program the instructions into the chip. The assembler software, MPASM, which converts your text to the machine code is available from Microchip on www.microchip.com – this website is a must for PIC programmers.

Program memory

Inside the microcontroller the program we write is stored in an area called EPROM (Electrically Programmable Read Only Memory). This memory is non-volatile and is remembered when the power is switched off. The memory is electrically programmed by a piece of hardware called a programmer.

The instructions we program into our microcontroller work by moving and manipulating data in memory locations known as user files and registers. This memory is called RAM, Random Access Memory. For example, in the room heater we would measure the room temperature by instructing the microcontroller via its Analogue to Digital Control Register (ADCON0), the measurement would then be compared with our data stored in one of the user files. A STATUS register would indicate if the temperature was above or below the required value and a PORT register would turn the heater on or off accordingly. The memory map of the 16F84 chip is shown in Chapter 16.

PIC microcontrollers are 8-bit micros, which means that the memory locations, the user files and registers are made up of 8 binary digits shown in Figure 1.1

bit 7	bit 6	bit 5	bit 4	bit 3	bit 2	bit 1	bit 0
1	0	1	1	0	0	1	0

MSB --LSB

Figure 1.1 User file and register layout

Bit 0 is the Least Significant Bit (LSB) and bit 7 is the Most Significant Bit (MSB). The use of these binary digits is explained in Appendix D.

When you make an analogue measurement, the digital number, which results, will be stored in a register called ADRES. If you are counting the number of times a light has been turned on and off, the result would be stored as an 8-bit binary number in a user file called, say, COUNT.

Microcontroller clock

In order to step through the instructions the microcontroller needs a clock frequency to orchestrate the movement of the data around its electronic circuits. This is usually provided by two capacitors and a crystal as shown in our first program example in Figure 2.1, page 15.

In the 16F84 microcontroller there are four oscillator options:

- An RC (Resistor/Capacitor) oscillator which provides a low cost solution.
- An LP oscillator, i.e. 32kHz crystal, which minimises power consumption.
- XT which uses a standard crystal configuration.
- HS which is the high-speed oscillator option.

Common crystal frequencies would be 32kHz, 1MHz, 4MHz, 10MHz and 20MHz.

Inside the microcontroller there is an area where the processing (the clever work), such as mathematical and logical operations, are performed; this is known as the central processing unit or CPU. There is also a region where event timing is performed and another for interfacing to the outside world through ports.

Selecting a microcontroller

In order to choose a microcontroller for a particular control system let us first of all consider the block diagram of the microcontroller system shown in Figure 1.2

Figure 1.2 The basic microcontroller system

- The input components would consist of digital devices such as switches, push buttons, pressure mats, float switches, keypads, radio receivers, etc. and analogue sensors such as light dependent resistors, thermistors, gas sensors, pressure sensors, etc.
- The control unit is of course the microcontroller. The microcontroller will monitor the inputs and as a result the program would turn outputs on and off. The microcontroller stores the program in its memory, and executes the instructions under the control of the clock circuit.
- The output devices would be made up from LEDs, buzzers, motors, alphanumeric displays, radio transmitters, 7-segment displays, heaters, fans, etc.

The most obvious choice then for the microcontroller is how many digital inputs, analogue inputs and outputs the system requires. This would then specify the minimum number of inputs and outputs (I/O) that the microcontroller must have. If analogue inputs are used then the microcontroller must have an Analogue to Digital (A/D) module inside.

The next consideration would be what size of program memory storage is required. This should not be too much of a problem when starting out, as most programs would be relatively small. All programs in this book fit into a 1k program memory space.

The clock frequency determines the speed at which the instructions are executed. This is important if any lengthy calculations are being undertaken. The higher the clock frequency the quicker the micro will finish one task and start another.

Other considerations are the number of interrupts and timer circuits required, and how much data EEPROM if any is needed. These more complex operations are considered later in the text.

In this book the programs requiring analogue inputs have been implemented on the 16C711 micro. Programs requiring only digital inputs have used the 16F84. Both of these devices have 1k of program memory and have been run using a 32.768kHz clock frequency. Even though there are over 100 PIC microcontrollers, the problem of which one to use need not be considered until you have understood the applications in this book and are ready to move onto some scary stuff.

Types of microcontroller

The list of PIC microcontrollers is growing almost daily. There are 105 at present and they include devices for all kinds of applications, for example the (planned) 18F852 will have 16k of EPROM memory, 2k bytes of RAM (user files), 16 12-bit A/D channels, a voltage reference, 68 inputs and outputs (I/O), three 16-bit and one 8-bit timers.

There are basically two types of microcontrollers, flash devices and One Time Programmable Devices (OTP). The flash devices can be reprogrammed in the programmer whereas OTP devices once programmed cannot be reprogrammed. All OTP devices, however, do have a windowed variety, which enables them to be erased under ultraviolet light in about 15 minutes, so that they can be reprogrammed. The windowed devices have a suffix JW to distinguish them from the others.

The OTP devices are specified for a particular oscillator configuration R-C, LP, XT or HS. See Appendix B Microcontroller data.

16C54 configurations are:

16C54JW	Windowed device
16C54RC	OTP, R-C oscillator
16C54LP	OTP, LP oscillator, 32kHz
16C54XT	OTP, XT oscillator, 4MHz
16C54HS	OTP, HS oscillator, 20MHz

In this book the two main devices investigated are the 16F84 flash device and the 16C711 OTP device with a four channel A/D. The 16F84 at present is the main choice for beginners. It has its program memory made using flash technology. It can be programmed, tested in a circuit and reprogrammed if required without the need for an ultraviolet eraser. The 16C711 is usually the beginners' choice for a device containing an A/D converter.

Microcontroller specification

A device is specified with its Product Identification Code. This code specifies:

- The device number.
- If it is a windowed, an OTP, or flash device. The windowed device is specified by a JW suffix. OTP devices are specified by oscillator frequency, and the flash devices are specified with an F such as 16F84.
- The oscillation frequency, usually 04 for devices working up to 4MHz, 10 up to 10MHz or 20 up to 20MHz. 20MHz devices are of course more expensive than 4MHz devices.
- Temperature range, for general applications 0°C to +70°C is usually specified.

The Product Identification System for the PIC micro is shown in Figure 1.3.

The devices used in this book are the 16F84-04/P, the 4MHz 16F84 in a plastic package, the 16C711JW, the windowed 16C711 device (windowed devices operate up to 10MHz) and the 12C508JW, the windowed (8 pin) 12C508.

Using the microcontroller

In order to use the microcontroller in a circuit there are basically two areas you need to understand:

1 how to connect the microcontroller to the hardware.
2 how to write and program the code into the microcontroller.

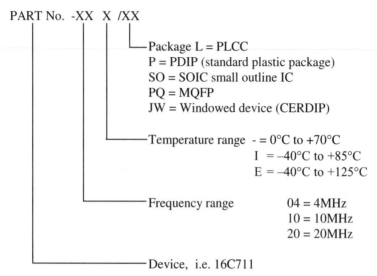

```
PART No.  -XX  X  /XX
```

Package L = PLCC
P = PDIP (standard plastic package)
SO = SOIC small outline IC
PQ = MQFP
JW = Windowed device (CERDIP)

Temperature range - = 0°C to +70°C
I = –40°C to +85°C
E = –40°C to +125°C

Frequency range   04 = 4MHz
10 = 10MHz
20 = 20MHz

Device, i.e. 16C711

Figure 1.3 Product Identification System

1 Microcontroller hardware

The hardware that the microcontroller needs to function is shown in Figure 1.4. The crystal and capacitors connected to pins 15 and 16 of the 16F84 produce the clock pulses that are required to step the microcontroller through the program and provide the timing pulses. The 0.1μF capacitor is placed as close to the chip as possible between 5v and 0v. Its role is to divert (filter) any electrical noise on the 5v power supply line to 0v, thus bypassing the microcontroller.

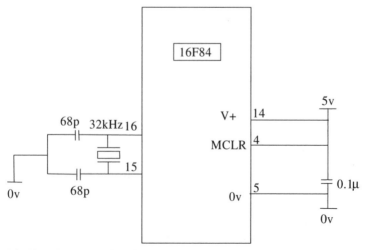

Figure 1.4 The microcontroller circuit

This capacitor must always be connected to stop any noise affecting the normal running of the microcontroller.

Microcontroller power supply

The power supply for the microcontroller needs to be between 2v and 6v. This can easily be provided from a 6v battery as shown in Figure 1.5.

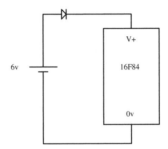

Figure 1.5 Microcontroller power supply

The diode in the circuit drops 0.7v across it reducing the applied voltage to 5.3v. It provides protection for the microcontroller if the battery is accidentally connected the wrong way round. In that case the diode would be reversed biased and no current would flow.

7805, voltage regulator circuit

Probably the most common power supply connection for the microcontroller is a three terminal voltage regulator IC, the 7805. The connection for this is shown in Figure 1.6.

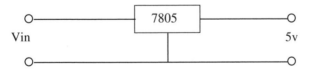

Figure 1.6 The voltage regulator circuit

The supply voltage, Vin, to the 7805 can be anything from 7v to 30v. The output voltage will be a fixed 5v and can supply currents up to 1 amp. So battery supplies such as 24v, 12v, 9v, etc. can be accommodated.

Power dissipation in the 7805

Care must be taken when using a high value for Vin. For example, if Vin = 24v the output of the 7805 will be 5v, so the 7805 has 24 − 5 = 19v across it. If it is supplying a current of 0.5 amp to the circuit then the power dissipated (volts × current) is 19 × 0.5 = 9.5 watts. The regulator will get hot and will need a heatsink to dissipate this heat.

If a supply of 9v is connected to the regulator it will have 4v across it and would dissipate 4 × 0.5 = 2 watts.

In the circuits used in this book the microcontroller only requires a current of 15µA so most of the current drawn will be from the outputs. If the output current is not too large, say <100mA (0.1A), then with a 9v supply the power dissipated would be 4 × 0.1 = 0.4 watts and the regulator will stay cool without a heatsink.

Connecting switches to the microcontroller

The most common way of connecting a switch to a microcontroller is via a pull-up resistor to 5v as shown in Figure 1.7.

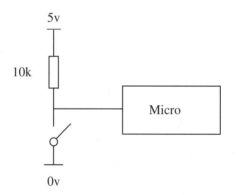

Figure 1.7 Connecting a switch to the Microcontroller

When the switch is open, 5v, a logic 1 is connected to the micro.
When the switch is closed, 0v, a logic 0 is connected to the micro.

Connecting outputs to the microcontroller

The microcontroller is capable of supplying approximately 20–25mA to an output pin. So loads such as LEDs or small relays can be driven directly. Larger loads require interfacing via a transistor for dc or a triac for ac. Opto-coupled devices provide an isolated interface between the microcontroller and the load.

2 Programming the microcontroller

In order to have the microcontroller perform some controlling action you need to communicate with it and tell it what those instructions are to be. When we communicate with one another we use a spoken language, when we communicate with a microcontroller we use a program language. The program language for the PIC microcontroller uses 35 words (instructions) in its vocabulary. A few more instructions are used in the bigger microcontrollers.

In order to communicate with the microcontroller we need to know what these 35 instructions are and how to use them. Not all 35 instructions are used in this book. In fact you can write meaningful programs using only five or six instructions.

2
Programming the 16F84 microcontroller

Microcontrollers are now providing us with a new way of designing circuits. Designs, which at one time required many digital ICs and lengthy Boolean algebra calculations, can now be programmed simply into one microcontroller. For example, a set of traffic lights would have required an oscillator circuit, counting and decoding circuits plus an assortment of logic gate ICs.

In order to use this exciting new technology we must learn how to program these microcontrollers.

The microcontroller I have chosen to start with is the 16F84-04/P, which means it is a flash device that can be electrically erased and reprogrammed without using an ultraviolet eraser. It can be used up to an oscillation frequency of 4MHz and comes in a standard 18 pin plastic package.

It has 35 instructions in its vocabulary, but like all languages not all of the instructions are used all of the time – you can go a long way on just a few. In order to teach you how to use these instructions I will be using a simple program to flash an LED on and off continually. This program introduces you to four instructions in five lines of code.

You are then encouraged to write your own program to flash two LEDs on and off alternately. The idea being, when you have understood my code you can then modify it for your own program, enhancing your understanding. Once you have written your first program you are ready to proceed. The book then continues with further applications such as traffic lights and disco lights to introduce more of the instructions that will increase your microcontroller vocabulary.

New instructions used in this chapter:

- BCF
- BSF
- CALL
- GOTO
- MOVLW
- MOVWF
- DECFSZ

Microcontroller inputs and outputs (I/O)

The microcontroller is a very versatile chip and can be programmed to operate in a number of different configurations. The 16F84 is a 13 I/O device, which means it has 13 inputs and outputs. The I/O can be configured in any combination, i.e. one input 12 outputs, six inputs seven outputs, or 13 outputs depending on your application. These I/O are connected to the outside world through registers called ports. The 16F84 has two ports, PORTA and PORTB. PORTA is a 5-bit port and has five I/O lines; PORTB has eight I/O.

Timing with the microcontroller

All microcontrollers have timer circuits onboard; some have four different timers. The 16F84 has one timer register called TIMER0. These timers run at a speed of ¼ of the clock speed. So if we use a 32 768Hz crystal the internal timer will run at ¼ of 32 768Hz, i.e. 8192Hz. If we want to turn an LED on for, say, 1 second we would need to count 8192 of these timing pulses. This is a lot of pulses! Fortunately within the microcontroller there is a register called an OPTION register, which allows us to slow down these pulses by a factor of 2, 4, 8, 16, 32, 64, 128 or 256. The OPTION register is discussed in Chapter 16. Setting the prescaler, as it is called, to divide by 256 in the OPTION register means that our timing pulses are now 8192/256 = 32Hz, i.e. 32 pulses a second. So to turn our LED on for 1 second we need only to count 32 pulses in TIMER0, or 16 for 0.5 seconds, or 160 for 5 seconds, etc.

Programming the microcontroller

In order to program the microcontroller we need to:

- Write the instructions in a program.
- Change the text into machine code that the microcontroller understands using a piece of software called an assembler.
- Blow the data into the chip using a programmer.

Let's consider the first task, writing the program. This can be done on any text editor. I prefer DOS EDIT because the lines are numbered enabling syntax errors to be located easily. Notepad is also a useful medium for this purpose, but unfortunately the lines are not numbered.

As you have seen above we need to configure the I/O and set the prescaler for the timing. If we do not set them the default conditions are that all PORT bits are inputs. A micro with no outputs is not much use! The default for the prescaler is that the clock rate is divided by 2. The program also needs to know what device it is intended for and also what the start address in the memory is.

If this is starting to sound confusing – do not worry, I have written a header program, which sets all the above conditions for you to use. These conditions can be changed later when you understand more about what you are doing.

The header sets the 5 bits of PORTA as inputs and the 8 bits of PORTB as outputs. It also sets the prescaler to divide by 256. We will use the 32 768Hz crystal so our timing is 32 pulses per second. The program instructions will run at ¼ of the 32 768Hz clock, i.e. 8192 instructions per second. The header also includes two timing subroutines for you to use: they are DELAY1 – a 1 second delay; and DELAYP5 – a half-second delay. A subroutine is a section of code that can be called, when needed, to save writing it again. For the moment do not worry about how the header or the delay subroutines work. We will work through them, in Chapter 3, once we have programmed a couple of applications.

Just one more point – there are different ways of entering data.

Entering data

Consider the decimal number 37, this has a hex value of 25 or a binary value of 0010 0101. The assembler will accept this as .37 in decimal (note the . is not a decimal point) or as 25H in hex or B'00100101' in binary.

181 decimal would be entered as .181 in decimal, 0B5H in hex or B'10110101' in binary. Note: If a hex number starts with a letter it must be prefixed with a 0, i.e. 0B5H not B5H. The default radix for the assembler MPASM is hex.

Appendix D illustrates how to change between decimal, binary and hexadecimal numbers.

The PIC microcontrollers are 8-bit micros. This means that the memory locations, i.e. user files and registers contain, 8 bits. So the smallest 8-bit number is of course 0000 0000 which is equal to a decimal number 0 (of course). The largest 8-bit number is 1111 1111 which is equal to a decimal number of 255. To use numbers larger than 255 we have to combine memory locations. Two memory locations combine to give 16 bits with numbers up to 65 536. Three memory locations combine to give 24 bits allowing numbers up to 16 777 215 and so on. These large numbers are introduced in Chapter 5, Numbers larger than 255.

The header for the 16F84, HEADER84.ASM

The listing below shows the header for the 16F84 microcontroller. I suggest you start all of your programs with this header, or a modified version of it. A full explanation of this header file is given in Chapter 3.

```
;HEADER84.ASM for 16F84.   This sets PORTA as an INPUT (NB 1
;                          means input) and PORTB as an OUTPUT;
;                          (NB 0 means output). The OPTION
;                          register is set to /256 to give timing pulses
;                          of 1/32 of a second.
;                          1 second and 0.5 second delays are
;                          included in the subroutine section.

;*********************************************************

;EQUATES SECTION

TMR0      EQU     1        ;means TMR0 is file 1.
STATUS    EQU     3        ;means STATUS is file 3.
PORTA     EQU     5        ;means PORTA is file 5.
PORTB     EQU     6        ;means PORTB is file 6.
ZEROBIT   EQU     2        ;means ZEROBIT is bit 2.
COUNT     EQU     0CH      ;means COUNT is file 0C,
                          ;a register to count events.
;*********************************************************

LIST      P=16F84          ;we are using the 16F84.
ORG       0                ;the start address in memory is 0
GOTO      START            ;goto start!

;*********************************************************

;SUBROUTINE SECTION.

;1 second delay.
DELAY1    CLRF    TMR0            ;START TMR0.
LOOPA     MOVF    TMR0,W          ;READ TMR0 INTO W.
          SUBLW   .32             ;TIME - 32
          BTFSS   STATUS,ZEROBIT  ;Check TIME-W = 0
          GOTO    LOOPA           ;Time is not = 32.
          RETLW   0               ;Time is 32, return.
```

;0.5 second delay.

```
DELAYP5    CLRF      TMR0                    ;START TMR0.
LOOPB      MOVF      TMR0,W                  ;READ TMR0 INTO W.
           SUBLW     .16                     ;TIME - 16
           BTFSS     STATUS,ZEROBIT          ;Check TIME-W = 0
           GOTO      LOOPB                   ;Time is not = 16.
           RETLW     0                       ;Time is 16, return.
```

;***
;

;CONFIGURATION SECTION

```
START      BSF       STATUS,5                ;Turns to Bank1.

           MOVLW     B'00011111'             ;5bits of PORTA are I/P
           TRIS      PORTA

           MOVLW     B'00000000'
           TRIS      PORTB                   ;PORTB is OUTPUT

           MOVLW     B'00000111'             ;Prescaler is /256
           OPTION                            ;TIMER is 1/32 secs.

           BCF       STATUS,5                ;Return to Bank0.
           CLRF      PORTA                   ;Clears PortA.
           CLRF      PORTB                   ;Clears PortB.
```

;***
;
;Program starts now.

```
END                  ;This must always come at the end of your code
```

Note: In the program any text on a line following the semicolon (;) is ignored by the assembler software. Program comments can then be placed there.

The section is saved as HEADER84.ASM – you can use it at the start of all your programs. HEADER84 is the name of our program and ASM is its extension, or you can save it as HEADER84.TXT.

Program example

The best way to begin to understand how to use a microcontroller is to start with a simple example and then build on this.

Let us consider a program to flash an LED ON and OFF at 0.5 second intervals. The LED is connected to PortB bit 0 as shown in Figure 2.1.

Notice from Figure 2.1 how few components the microcontroller needs – 2 × 68pF capacitors, a 32.768kHz crystal for the oscillator and a 0.1µF capacitor for decoupling the power supply. Other oscillator and crystal configurations are possible – see Microchip's data sheets for other combinations. I have chosen the 32kHz crystal because it enables times of seconds to be produced easily.

The program for this circuit can be written on any text editor, probably the most useful is the DOS EDIT. If you prefer you can use any text editor, e.g. Notepad.

- In EDIT. If you have HEADER84.ASM on disk type A:\ EDIT HEADER84.ASM enter, to load the file. If you do not have it on disk type A:\ EDIT enter, and enter the text and SAVE AS HEADER84.ASM.
- In Notepad open HEADER84.ASM or start a new file and type the program in, saving as HEADER84.ASM as type 'All Files' to avoid Notepad adding the extension .TXT.

Once you have HEADER84.ASM saved on disk and loaded onto the screen alter it by including your program as shown below:

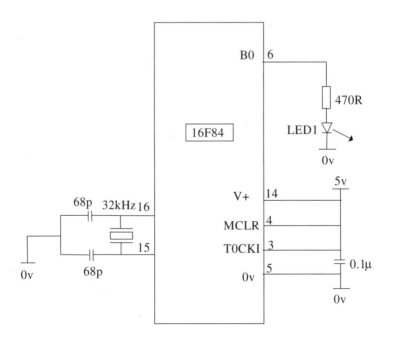

Figure 2.1 Circuit diagram of the microcontroller flasher

```
;HEADER84.ASM for 16F84.  This sets PORTA as an INPUT (NB 1
;                         means input) and PORTB as an OUTPUT
;                         (NB 0 means output). The OPTION
;                         register is set to /256 to give timing pulses
;                         of 1/32 of a second.
;                         1 second and 0.5 second delays are
;                         included in the subroutine section.
;************************************************************

;EQUATES SECTION

TMR0        EQU     1                       ;means TMR0 is file 1.
STATUS      EQU     3                       ;means STATUS is file 3.
PORTA       EQU     5                       ;means PORTA is file 5.
PORTB       EQU     6                       ;means PORTB is file 6.
ZEROBIT     EQU     2                       ;means ZEROBIT is bit 2.
COUNT       EQU     0CH                     ;means COUNT is file 0C,
                                            ;a register to count events.
;************************************************************

LIST        P=16F84 ;we are using the 16F84.
ORG         0       ;the start address in memory is 0
GOTO        START   ;goto start!

;************************************************************

;SUBROUTINE SECTION.

;1 second delay.
DELAY1      CLRF    TMR0                    ;START TMR0.
LOOPA       MOVF    TMR0,W                  ;READ TMR0 INTO W.
            SUBLW   .32                     ;TIME - 32
            BTFSS   STATUS,ZEROBIT          ;Check TIME-W = 0
            GOTO    LOOPA                   ;Time is not = 32.
            RETLW   0                       ;Time is 32, return.


;0.5 second delay.
DELAYP5     CLRF    TMR0                    ;START TMR0.
LOOPB       MOVF    TMR0,W                  ;READ TMR0 INTO W.
            SUBLW   .16                     ;TIME - 16
            BTFSS   STATUS,ZEROBIT          ;Check TIME-W = 0
            GOTO    LOOPB                   ;Time is not = 16.
            RETLW   0                       ;Time is 16, return.

;************************************************************
```

;CONFIGURATION SECTION

START	BSF	STATUS,5	;Turns to Bank1.
	MOVLW	B'00011111'	;5 bits of PORTA are I/P
	TRIS	PORTA	
	MOVLW	B'00000000'	
	TRIS	PORTB	;PORTB is OUTPUT
	MOVLW	B'00000111'	;Prescaler is /256
	OPTION		;TIMER is 1/32 secs.
	BCF	STATUS,5	;Return to Bank0.
	CLRF	PORTA	;Clears PortA.
	CLRF	PORTB	;Clears PortB.

;***
;Program starts now.

BEGIN	BSF	PORTB,0	;Turn ON B0.
	CALL	DELAYP5	;Wait 0.5 seconds
	BCF	PORTB,0	;Turn OFF B0.
	CALL	DELAYP5	;Wait 0.5 seconds
	GOTO	BEGIN	;Repeat
END			;YOU MUST END!!

How does it work?

The five lines of code starting at BEGIN are responsible for flashing the LED ON and OFF. This is all the code we will require for now. The rest of the code, the header, is explained in Chapter 3 once you have seen the program working.

• BEGIN is a label. A label is used as a location for the program to go to.
• Line1 the instruction BSF and its data PORTB,0 is shorthand for Bit Set in File, which means Set the Bit in the File PORTB, where bit0 is the designated bit. This will cause PORTB,0 to be Set to a logic1, in hardware terms this means pin6 in Figure 2.1 is at 5v turning the LED on.

Note: There must not be any spaces in a label, an instruction or its data. I keep the program tidy by using the TAB key on the keyboard.

• Line2 CALL DELAYP5 causes the program to wait 0.5 seconds while the subroutine DELAYP5 in the header is executed.
• Line3 BCF PORTB,0 is the opposite of Line1, this code is shorthand for Bit Clear in File, which means Clear the Bit in the File PORTB, where bit0 is the designated bit. This will cause PORTB,0 to be Cleared to a logic0, in hardware terms this means pin6 in Figure 2.1 is at 0v turning the LED off.

- Line4 CALL DELAYP5 is the same as Line2.
- Line5 GOTO BEGIN sends the program back to the label BEGIN to repeat the process of flashing the LED on and off.

Any of the eight outputs can be turned ON and OFF using the two instructions BSF and BCF, for example:

BSF PORTB,3 makes PORTB,3 (pin 9) 5v.
BCF PORTB,7 makes PORTB,7 (pin 13) 0v.

Saving and assembling the program

The program is then saved as FLASHER.ASM. Exit the EDIT program (or Notepad, etc.). The next task is to assemble this text into the hex code that the microcontroller understands. This is done by executing the assembler program, MPASMWIN, available on the Internet from Microchip.com. Open the program MPASMWIN. Screen 2.1 will open up.

Type in the source file name A:\Flasher.ASM or click Browse and select it from your selected drive and directory, i.e. C:\Micro. Click Assemble, the software will then change your text into the machine code and generate an error file, FLASHER.ERR, a listing file, FLASHER.LST and a hex file, FLASHER.HEX.

The error file will list the syntax errors in your FLASHER.ASM file, hopefully there will not be any!
The listing file will list your program in detail giving line numbers and object codes and also a more detailed error report than was given on the error file.
The hex file is the file used to program the code into the chip.
To view these files, e.g. FLASHER.LST, Type:
EDIT FLASHER.LST if using EDIT
Or to view the errors type:
EDIT FLASHER.ERR
Or to view the hex file type:
EDIT FLASHER.HEX

Or view with Notepad, by opening the required file.

If you have any errors then view the error file and correct the errors in your file FLASHER.ASM. These errors are syntax errors, i.e. spelling mistakes or perhaps commas have been missed out. Resave FLASHER.ASM and repeat the assembly-error correction process until all errors are cleared. The assembler

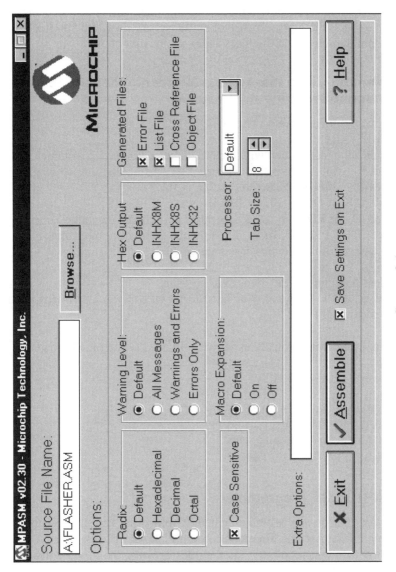

Screen 2.1

software will then have generated the file FLASHER.HEX that is used to blow the program into the microcontroller.

Programming the code into the microcontroller

There are a number of programmers on the market. Some are dedicated machines, others (the more expensive) are universal. Basically you load your hex file onto the screen, set the fuses and program your chip.

Programming using the Arizona Microchip Picstart 16B1 Development System

Connect the programmer and switch on.
Type: MPSTART
After establishing connection with the programmer Screen 2.2 will open up.

Select Device – 16F84. The default setting is 16C54.
Select File, ALT-F, OPEN, FLASHER.HEX.
Next select FUSE EDIT and select the fuses, switch the Watchdog Timer OFF and select CODE PROTECT ON if you require your code protecting.
Press F5 and your microcontroller is programmed.

Note: Most programs in this book will be using the 16F84 chip, with an LP oscillator, i.e. a 32kHz crystal, Watchdog Timer OFF, Power Up Timer ON and Code Protection OFF.

Whichever programmer you are using you will need to enter the following:

- the File, i.e. FLASHER.HEX
- Device type, i.e. 16F84
- the Oscillator, RC, LP, XT, HS, i.e. LP
- Watchdog Timer ON/OFF, i.e. OFF
- Power Up Timer ON/OFF, i.e. ON
- Code Protect ON/OFF, i.e. OFF

PICSTART PLUS programmer

The Picstart 16B1 is now a little outdated. If you do not have a programmer I would recommend Arizona Microchip's own PICSTART PLUS. When Arizona bring out a new microcontroller, as they do regularly, the driver software is updated and can be downloaded free off the Internet from microchip.com. Once installed on your PC it is opened from MPLAB, i.e.

```
PicStart File Windows Options Device Transfer Config  P16C84  22:21:08

          ┌─────── C:\MICRO\FLASHER.HEX ───────┐        ┌ [ ] Fuses ──────┐
0000:  280D 0181 0801 3C20 1D03 2802 3400 0181  .. ....│ Osc: LP
0008:  0801 3C10 1D03 2808 3400 1683 301F 0065  ......e │ WDT: Off
0010:  3000 0066 3007 0062 1283 0185 0186 1406  .f.b....│ PuT: On
0018:  2007 1006 2007 2817 3FFF 3FFF 3FFF 3FFF  ........│ CP : Off
0020:  3FFF 3FFF 3FFF 3FFF 3FFF 3FFF 3FFF 3FFF  ........│ ID : 7F7F7F7F
0028:  3FFF 3FFF 3FFF 3FFF 3FFF 3FFF 3FFF 3FFF  ........│ CkSum: 2AFA
0030:  3FFF 3FFF 3FFF 3FFF 3FFF 3FFF 3FFF 3FFF  ........│
0038:  3FFF 3FFF 3FFF 3FFF 3FFF 3FFF 3FFF 3FFF  ........│     Id Edit
0040:  3FFF 3FFF 3FFF 3FFF 3FFF 3FFF 3FFF 3FFF  ........│
0048:  3FFF 3FFF 3FFF 3FFF 3FFF 3FFF 3FFF 3FFF  ........│    Fuse Edit
0050:  3FFF 3FFF 3FFF 3FFF 3FFF 3FFF 3FFF 3FFF  ........│
0058:  3FFF 3FFF 3FFF 3FFF 3FFF 3FFF 3FFF 3FFF  ........└─────────────────┘
0060:  3FFF 3FFF 3FFF 3FFF 3FFF 3FFF 3FFF 3FFF  ........
0068:  3FFF 3FFF 3FFF 3FFF 3FFF 3FFF 3FFF 3FFF  ........
0070:  3FFF 3FFF 3FFF 3FFF 3FFF 3FFF 3FFF 3FFF  ........

F1 Help  Alt-X Exit    F4 Edit  F5 Program  F6 Verify  F7 Blank  F8 Read
```

Screen 2.2

Open MPLAB
Click Picstart Plus (from the MENU bar)
Click Enable Programmer
A screen similar to Figure 2.2 will open up
Select the device PIC16F84 from the list
Click File, select Import, Download To Memory
Load your hex file, i.e. A:\FLASHER.HEX

Select	Oscillator	LP
	Watchdog Timer	Off
	Power Up Timer	On
	Code Protect	Off (no need to protect this just yet!)

Your screen should now look like Figure 2.2

Click Program (to program the micro of course). After a short while the message 'success' will appear on the screen. You will be greeted with the success statement for a few seconds only, if you miss it check the program statistics for

Programming flowchart

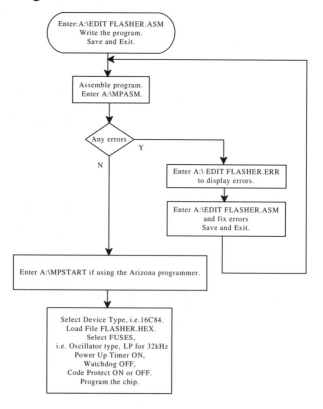

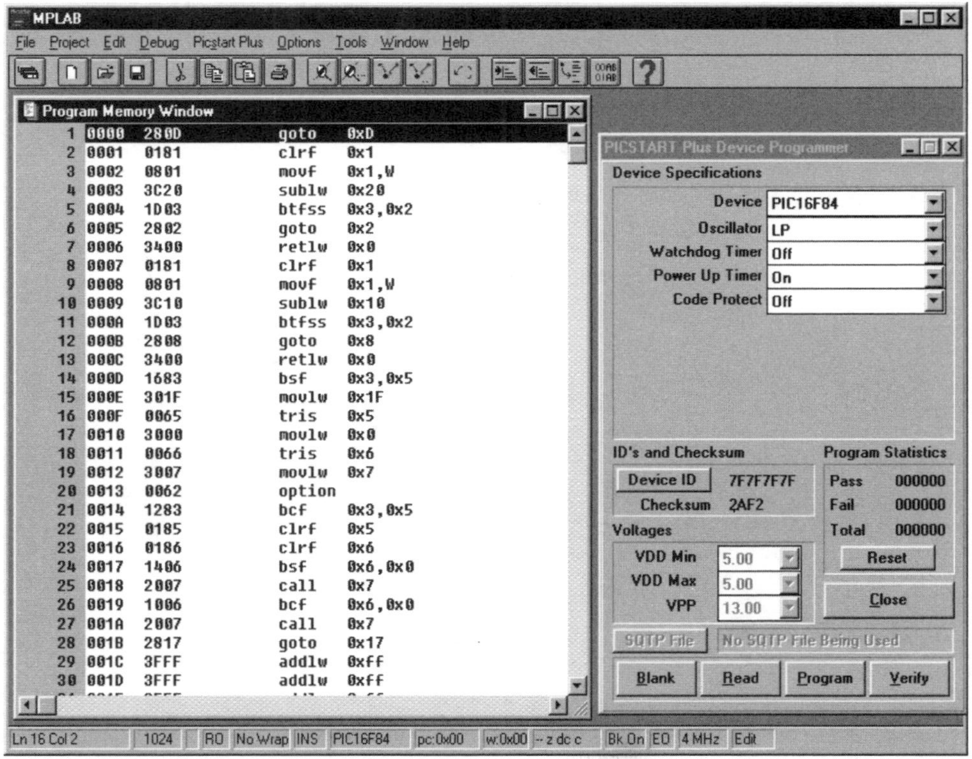

Figure 2.2 MPLAB screen

Pass 1 Fail 0 Total 1, which will be continually updated. The code has been successfully blown into your chip and is ready for use. If this process fails – check the chip is inserted correctly in the socket; if it is then try another chip.

So we are now able to use the microcontroller to switch an LED on and off – fantastic! But use your imagination. There are 35 instructions in your micro vocabulary. The PIC microcontroller range at the moment includes devices with 8k of EPROM program memory, 454 × 8 bits of RAM data memory, 33 input and output pins, 11 interrupts, eight channel A/D converter, 20MHz clock, real time clock, four counter/timers, 55 word instruction set – See Appendix B for a detailed list. If the 8k of EPROM or 454 bytes of RAM is not enough your system can be expanded using extra EPROM and RAM. In the end the only real limits will be your imagination.

Problem: flashing two LEDs

There has been a lot to do and think about to get this first program into the microcontroller and make it work in a circuit. But just so that you are sure of what you are doing – write a program that will flash two LEDs on and off alternately. Put LED0 on B0 and LED1 on B1. Note: You can use the file FLASHER.ASM, it only needs two extra lines adding! Then save it as FLASHER2.ASM. Try not to look at the solution below before you have attempted it.

Solution to the problem: flashing two LEDs
The header is the same as in FLASHER.ASM. Just include in the section Program starts now the following lines:

;Program starts now.

```
BEGIN      BSF       PORTB,0       ;Turn ON B0.
           BCF       PORTB,1       ;Turn OFF B1
           CALL      DELAYP5       ;Wait 0.5 seconds
           BCF       PORTB,0       ;Turn OFF B0.
           BSF       PORTB,1       ;Turn ON B1.
           CALL      DELAYP5       ;Wait 0.5 seconds
           GOTO      BEGIN         ;Repeat
END
```

Did you manage to do this? If not have a look at my solution and see what the lines are doing. If you did not manage this, try flashing four LEDs on and off, with two on and two off alternately. You might like to have them on for 1 second and off for half a second. Can you see how to use the 1-second delay in place of the half-second delay? The different combinations of switching any eight LEDs on PORTB should be relatively easy once you have mastered these steps. Consider another example of the delay routine.

Traffic lights

If you have ever tried to design a 'simple' set of traffic lights then you will appreciate how much circuitry is required: an oscillator circuit, counters and logic decode circuitry. The microcontroller circuit is a much better solution even for this 'simple' arrangement. The circuit is shown in Figure 2.3.

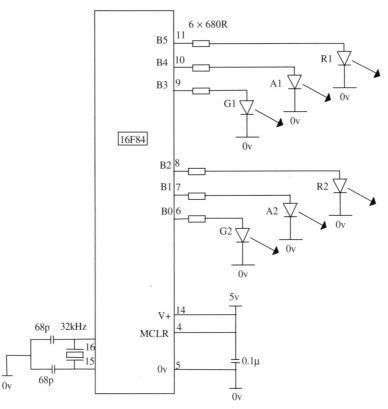

Figure 2.3 Traffic lights circuit

A truth table of the operation of the lights is probably a better aid to a solution rather than a flowchart.

Traffic lights truth table

Time	B7	B6	B5	B4	B3	B2	B1	B0
			R1	A1	G1	R2	A2	G2
2 sec	0	0	1	0	0	1	0	0
2 sec	0	0	1	1	0	1	0	0
5 sec	0	0	0	0	1	1	0	0
2 sec	0	0	0	1	0	1	0	0
2 sec	0	0	1	0	0	1	0	0
2 sec	0	0	1	0	0	1	1	0
5 sec	0	0	1	0	0	0	0	1
2 sec	0	0	1	0	0	0	1	0
				REPEAT				

Program listing for the traffic lights

;TRAFFIC.ASM

;EQUATES SECTION

```
TMR0      EQU     1           ;means TMR0 is file 1.
STATUS    EQU     3           ;means STATUS is file 3.
PORTA     EQU     5           ;means PORTA is file 5.
PORTB     EQU     6           ;means PORTB is file 6.
ZEROBIT   EQU     2           ;means ZEROBIT is bit 2.
COUNT     EQU     0CH         ;means COUNT is file 0C,
                              ;a register to count events.
;****************************************************
LIST      P=16F84   ;we are using the 16F84.
ORG       0         ;the start address in memory is 0
GOTO      START     ;goto start!

;****************************************************
;SUBROUTINE SECTION.

;5 second delay.
DELAY5    CLRF    TMR0        ;Start TMR0.
LOOPA     MOVF    TMR0,W      ;Read TMR0 intoW.
          SUBLW   .160        ;TIME - 160
          BTFSS   STATUS,ZEROBIT ; Check TIME-W = 0
          GOTO    LOOPA       ;Time is not = 160.
          RETLW   0           ;Time is 160, return.
```

```
;2 second delay.
DELAY2    CLRF    TMR0              ;Start TMR0.
LOOPB     MOVF    TMR0,W            ;Read TMR0 into W.
          SUBLW   .64               ;TIME - 64
          BTFSS   STATUS,ZEROBIT    ; Check TIME-W = 0
          GOTO    LOOPB             ;Time is not = 64.
          RETLW   0                 ;Time is 64, return.
;********************************************************
;CONFIGURATION SECTION

START     BSF     STATUS,5          ;Turns to Bank1.
          MOVLW   B'00011111'       ;5bits of PORTA are I/P
          TRIS    PORTA
          MOVLW   B'00000000'
          TRIS    PORTB             ;PORTB is an output port.
          MOVLW   B'00000111'       ;Prescaler is /256
          OPTION                    ;TIMER is 1/32 secs.
          BCF     STATUS,5          ;Return to Bank0.
          CLRF    PORTA             ;Clears PortA.
          CLRF    PORTB             ;Clears PortB.
;********************************************************
;Program starts now.

BEGIN     MOVLW   B'00100100'       ;R1, R2 on.
          MOVWF   PORTB
          CALL    DELAY2            ;Wait 2 Seconds.

          MOVLW   B'00110100'       ;R1, A1, R2 on.
          MOVWF   PORTB
          CALL    DELAY2            ;Wait 2 Seconds.

          MOVLW   B'00001100'       ;G1, R2 on.
          MOVWF   PORTB
          CALL    DELAY5            ;Wait 5 Seconds.

          MOVLW   B'00010100'       ;A1, R2 on.
          MOVWF   PORTB
          CALL    DELAY2            ;Wait 2 Seconds.

          MOVLW   B'00100100'       ;R1, R2 on.
          MOVWF   PORTB
          CALL    DELAY2            ;Wait 2 Seconds.
```

```
              MOVLW      B'00100110'      ;R1, R2, A2 on.
              MOVWF      PORTB
              CALL       DELAY2           ;Wait 2 Seconds.

              MOVLW      B'00100001'      ;R1, G2 on.
              MOVWF      PORTB
              CALL       DELAY5           ;Wait 5 Seconds.

              MOVLW      B'00100010'      ;R1, A2 on.
              MOVWF      PORTB
              CALL       DELAY2           ;Wait 2 Seconds.
              GOTO       BEGIN
END
```

How does it work?

In the previous example, FLASHER.ASM, we turned one LED on and off. We could have used the two commands BSF and BCF to turn several outputs on and off, but a much better way has been used with the TRAFFIC.ASM program.

The basic difference is the introduction of two more commands:

- MOVLW   MOVe the Literal (a number) into the Working register.
- MOVWF   MOVe the Working register to the File.

The data in this example, binary numbers, are moved to W and then to the file which is the output PORTB to switch the LEDs on and off. Unfortunately the data cannot be placed in PORTB with only one instruction; it has to go via the W register. So:

MOVLW   B'00100100'   clears B7, B6, sets B5, clears B4, B3, sets
                            B2 and clears B1, B0 in the W register
MOVWF   PORTB   moves the data from the W register to
                            PORTB to turn the relevant LEDs on and
                            off.

All eight outputs are turned on/off with these two instructions.

CALL DELAY2   waits 2 seconds before continuing with the
                            next operation. DELAY2 has been added to
                            the subroutine section.

The W register

The W or Working register is the most important register in the micro. It is in the W register where all the calculations and logical manipulations such as addition, subtraction, and-ing, or-ing, etc. are done.

The W register shunts data around like a telephone exchange reroutes telephone calls. In order to move data from locationA to locationB, the data has to be moved from locationA to W and then from W to locationB.

Note: If the three lines in the TRAFFIC.ASM program are repeated then any pattern and any delay can be used to sequence the lights – you can make your own disco lights!

Repetition (e.g. disco lights)

Instead of just repeating one sequence over and over, suppose we wish to repeat several sequences before returning to the start as with a set of disco lights.

Consider the circuit shown in Figure 2.4. The eight 'Disco Lights' B0–B7 are to be run as two sequences.

Sequence 1	Turn all lights on
	Wait
	Turn all lights off
	Wait
Sequence 2	Turn B7–B4 ON, B3–B0 OFF
	Wait
	Turn B7–B4 OFF, B3–B0 ON
	Wait

Suppose we wish Sequence 1 to run 5 times before going onto Sequence 2 to run 10 times and then repeat. A section of program is repeated a number of times with four lines of code shown below:

```
MOVLW    .5         ;Move 5 into W
MOVWF    COUNT      ;Move W into user file COUNT
.
SEQ1
.
DECFSZ   COUNT      ;decrement file COUNT skip if zero.
GOTO     SEQ1       ;COUNT not yet zero, repeat sequence
```

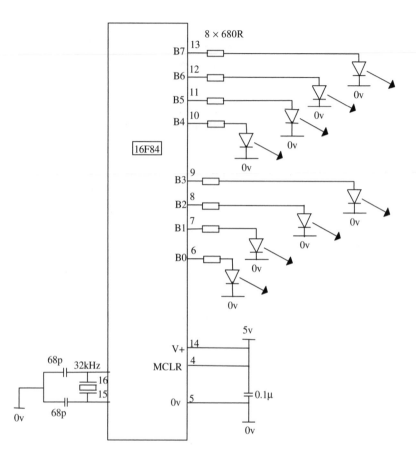

Figure 2.4 Disco lights

- The first two lines set up a file COUNT with 5. (Count is the first user file and is found in memory location 0CH.) 5 is first moved into W then from there to file COUNT.
- SEQ1 is executed.
- The DECFSZ COUNT instruction, DECrement File and Skip if Zero, decrements, takes 1 off the file COUNT and skips GOTO SEQ1 if the count is zero, if not zero then do SEQ1 again.

This way SEQ1 is executed five times and COUNT goes from 5 to 4 to 3 to 2 to 1 to 0 when we skip and follow onto SEQ2. SEQ2 is then done 10 times, say, and the code would be:

```
MOVLW      .10         ;Move 10 into W
MOVWF      COUNT       ;Move W into user file COUNT
.
SEQ2
.
DECFSZ     COUNT       ;decrement file COUNT skip if zero.
GOTO       SEQ2 ;      COUNT not yet zero, repeat sequence
```

Program code for the disco lights

```
;DISCO.ASM

;EQUATES SECTION

TMR0       EQU    1        ;means TMR0 is file 1.
STATUS     EQU    3        ;means STATUS is file 3.
PORTA      EQU    5        ;means PORTA is file 5.
PORTB      EQU    6        ;means PORTB is file 6.
ZEROBIT    EQU    2        ;means ZEROBIT is bit 2.
COUNT      EQU    0CH      ;means COUNT is file 0C,
                          ;a register to count events.
;*********************************************************
LIST       P=16F84         ;we are using the 16F84.
ORG        0               ;the start address in memory is 0
GOTO       START           ;goto start!

;*********************************************************
;SUBROUTINE SECTION.

;0. 5 second delay.
DELAYP5    CLRF   TMR0            ;Start TMR0.
LOOPA      MOVF   TMR0,W          ;Read TMR0 into W.
           SUBLW  .16             ;TIME - 16
           BTFSS  STATUS,ZEROBIT  ;Check TIME-W = 0
           GOTO   LOOPA           ;Time is not = 16.
           RETLW  0               ;Time is 16, return.

;*********************************************************

;CONFIGURATION SECTION
START      BSF    STATUS,5         ;Turns to Bank1.
           MOVLW  B'00011111'      ;5 bits of PORTA are I/P
           TRIS   PORTA
           MOVLW  B'00000000'
```

```
                TRIS      PORTB              ;PORTB is OUTPUT
                MOVLW     B'00000111'        ;Prescaler is /256
                OPTION                       ;TIMER is 1/32 secs.
                BCF       STATUS,5           ;Return to Bank0.
                CLRF      PORTA              ;Clears PortA.
                CLRF      PORTB              ;Clears PortB.
;*********************************************************
;Program starts now.

BEGIN           MOVLW     .5
                MOVWF     COUNT              ;Set COUNT = 5

SEQ1            MOVLW     B'11111111'
                MOVWF     PORTB              ;Turn B7-B0 ON
                CALL      DELAYP5            ;Wait 0.5 seconds
                MOVLW     B'00000000'
                MOVWF     PORTB              ;Turn B7-B0 OFF
                CALL      DELAYP5            ;Wait 0.5 seconds

                DECFSZ    COUNT              ;COUNT-1, skip if 0.
                GOTO      SEQ1

                MOVLW     .10
                MOVWF     COUNT              ;Set COUNT = 10

SEQ2            MOVLW     B'11110000'
                MOVWF     PORTB              ;B7-B4 on, B3-B0 off
                CALL      DELAYP5            ;Wait 0.5 seconds
                MOVLW     B'00001111'
                MOVWF     PORTB              ;B7-B4 off, B3-B0 on
                CALL      DELAYP5            ;Wait 0.5 seconds

                DECFSZ    COUNT              ;COUNT-1, skip if 0.
                GOTO      SEQ2
                GOTO      BEGIN
END
```

Using the idea of repeating sequences like this any number of combinations can be repeated. The times of course do not need to be of 0.5 seconds duration. The flash rate can be speeded up or slowed down depending on the combination. Try programming a set of your own disco lights. This should keep you quiet for hours (days!).

More than eight outputs

Suppose we wish to have a set of disco lights in a 3 x 3 matrix as shown in Figure 2.5. This configuration of course requires nine outputs. We have eight outputs on PORTB so we need to make one of the PORTA bits an output also, say PORTA bit0.

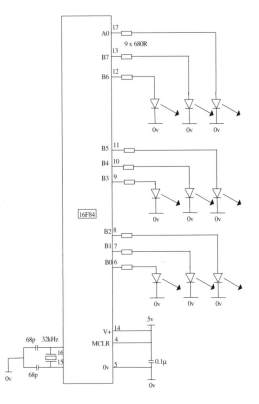

Figure 2.5 Nine disco light set

To change PORTA bit0 from an input to an output change the lines in the Configuration Section from:

```
MOVLW      B'00011111'
TRIS       PORTA
to
MOVLW      B'00011110'
TRIS       PORTA
```

Note: A 1 signifies an input, a 0 signifies an output.

So to set a '+' pattern in the lights we turn on B7, B4, B1, B3 and B5, keeping the others off. The code for this would be:

```
MOVLW       B'00000000'
MOVWF       PORTA              ;A0 is clear
MOVLW       B'10111010'
MOVWF       PORTB              ;B7, B5, B4, B3 and B1 are on
```

So to set an 'X' pattern in the lights we turn on B6, B4, B2, A0 and B0, keeping the others off. The code for this would be:

```
MOVLW       B'00000001'
MOVWF       PORTA              ;A0 is on
MOVLW       B'01010101'
MOVWF       PORTB              ;B6, B4, B2 and B0 are on
```

There are endless combinations you can make with nine lights. In fact there are 512. That is 2^9.

3
Using inputs

A control program usually requires more than turning outputs on and off. They switch on and off because an event has happened. This event is then connected to the input of the microcontroller to 'tell' it what to do next. The input could be derived from a switch or it could come from a sensor measuring temperature, light levels, soil moisture, air quality, fluid pressure, engine speed, etc.

Analogue inputs are dealt with later; in this chapter we will concern ourselves with digital on/off inputs.

New instructions used in this chapter:

- BTFSC
- BTFSS
- CLRF
- MOVF
- SUBLW
- SUBWF
- RETLW
- TRIS
- OPTION

As an example let us design a circuit so that switch SW1 will turn an LED on and off. The circuit diagram is shown in Figure 3.1.

The program to control the hardware would use the following steps:

1. Wait for SW1 to close.
2. Turn on LED1.
3. Wait for SW1 to open.
4. Turn off LED1.
5. Repeat.

In the circuit diagram SW1 is connected to A0 and LED1 to B0.
When the switch is closed A0 goes low or clear. So we wait until A0 is clear. The code for this is:

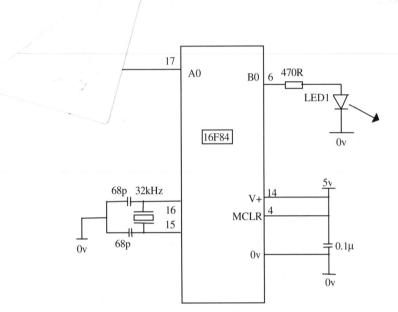

Figure 3.1 Circuit diagram of the microcontroller switch

```
BEGIN    BTFSC          PORTA,0 (test bit 0 in file PORTA skip if clear)
         GOTO           BEGIN
         BSF            PORTB,0
```

- The command BTFSC is Bit Test in File Skip if Clear, and the instruction BTFSC PORTA,0 means Test the Bit in the File PORTA, i.e. bit 0, Skip the next instruction if Clear. If A0 is Clear Skip the next instruction (GOTO BEGIN), if it isn't Clear then do not Skip and GOTO BEGIN to check the switch again.

The program will check the switch thousands maybe millions of times a second, depending on your clock.

- When the switch is pressed the program moves on and executes the instruction BSF PORTB,0 to turn on the LED.

We then wait for the switch to open. When the switch is open A0 goes Hi or Set. We then wait until A0 is Set, i.e.

```
SWOFF    BTFSS          PORTA,0
         GOTO           SWOFF
         BCF            PORTB,0
         GOTO           BEGIN
```

- The command BTFSS is Bit Test in File Skip if Set, and the instruction BTFSS PORTA,0 means Test the Bit in the File PORTA, i.e. bit 0, Skip the next instruction if Set. If A0 is Set Skip the next instruction (GOTO SWOFF), if it isn't Set then do not Skip and GOTO SWOFF to check the switch again.
- When the switch is set the program moves on and executes the instruction BCF PORTB,0 to switch off the LED.
- The program then goes back to the label BEGIN, to repeat.

The program is now added to the header. Note: Use the TAB to make your listing easy to read. The header has been altered to remove some of the unwanted items. It is then saved as SWITCH.ASM.

SWITCH.ASM

;SWITCH.ASM

;EQUATES SECTION

```
TMR0      EQU      1       ;means TMR0 is file 1.
STATUS    EQU      3       ;means STATUS is file 3.
PORTA     EQU      5       ;means PORTA is file 5.
PORTB     EQU      6       ;means PORTB is file 6.
ZEROBIT   EQU      2       ;means ZEROBIT is bit 2.
COUNT     EQU      0CH     ;means COUNT is file 0C,
                          ;a register to count events.
;*********************************************************

LIST      P=16F84          ;we are using the 16F84.
ORG       0                ;the start address in memory is 0
GOTO      START            ;goto start!

;*********************************************************
;CONFIGURATION SECTION

START     BSF      STATUS,5      ;Turns to Bank1.
          MOVLW    B'00011111'   ;5 bits of PORTA are I/P
          TRIS     PORTA
          MOVLW    B'00000000'
          TRIS     PORTB         ;PORTB is OUTPUT
          MOVLW    B'00000111'   ;Prescaler is /256
          OPTION                 ;TIMER is 1/32 secs.
          BCF      STATUS,5      ;Return to Bank0.
          CLRF     PORTA         ;Clears PortA.
```

```
            CLRF         PORTB        ;Clears PortB.

;************************************************************
;Program starts now.

BEGIN       BTFSC        PORTA,0      ;Wait for SW1 to be pressed
            GOTO         BEGIN
            BSF          PORTB,0      ;Turn on LED1.
SWOFF       BTFSS        PORTA,0      ;Wait for SW1 to be released.
            GOTO         SWOFF
            BCF          PORTB,0      ;Switch off LED1.
            GOTO         BEGIN        ;Repeat sequence.

END
```

Switch flowchart

It will be obvious from the program listing of the solution to the switch problem that listings are difficult to follow. A picture is worth a thousand words has never been more apt than it is with a program listing. The picture of the program is shown in Figure 3.2, the flowchart for the solution to our initial switch problem. Before a programming listing is attempted it is very worthwhile drawing a flowchart to depict the program steps. Diamonds are used to show a decision (i.e. a branch) and rectangles are used to show a command. Each shape may take several lines of program to implement. But the idea of the flowchart should be evident. Note that the flowchart describes the problem – you can draw it without any knowledge of the instruction set.

Program development

From our basic switch circuit an obvious addition would be to include a delay so that the LED would go off automatically after a set time. Suppose we wish to switch the light on for 5 seconds, using A0 as the switch input.

The flowchart for this delay is shown in Figure 3.3.

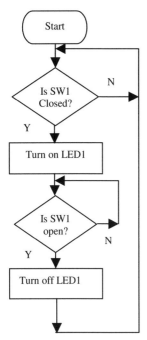

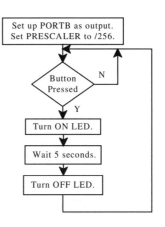

Figure 3.2 Flowchart for switch

Figure 3.3 Delay flowchart

The complete listing for the program for the 16F84

;DELAY.ASM

;EQUATES SECTION

TMR0	EQU	1	;means TMR0 is file 1.
STATUS	EQU	3	;means STATUS is file 3.
PORTA	EQU	5	;means PORTA is file 5.
PORTB	EQU	6	;means PORTB is file 6.
ZEROBIT	EQU	2	;means ZEROBIT is bit 2.
COUNT	EQU	0CH	;means COUNT is file 0C, ;a register to count events.

;**
;
LIST	P=16F84	;we are using the 16F84.
ORG	0	;the start address in memory is 0
GOTO	START	;goto start!

;**
;

;SUBROUTINE SECTION.

;5 second delay.

```
DELAY5      CLRF      TMR0               ;Start TMR0.
LOOPA       MOVF      TMR0,W             ;Read TMR0 into W.
            SUBLW     .160               ;TIME – 160
            BTFSS     STATUS,ZEROBIT     ;Check TIME-W = 0
            GOTO      LOOPA              ;Time is not = 160.
            RETLW     0                  ;Time is 160, return.
```

;***

;CONFIGURATION SECTION

```
START       BSF       STATUS,5           ;Turns to Bank1.
            MOVLW     B'00011111'        ;5 bits of PORTA are I/P
            TRIS      PORTA
            MOVLW     B'00000000'
            TRIS      PORTB              ;PORTB is OUTPUT
            MOVLW     B'00000111'        ;Prescaler is /256
            OPTION                       ;TIMER is 1/32 secs.
            BCF       STATUS,5           ;Return to Bank0.
            CLRF      PORTA              ;Clears PortA.
            CLRF      PORTB              ;Clears PortB.
```
;***
;Program starts now.

```
ON          BTFSC     PORTA,0            ;Check button pressed.
            GOTO      ON
            BSF       PORTB,0            ;Turn on LED.
            CALL      DELAY5             ;CALL 5 second delay
            BCF       PORTB,0            ;Turn off LED.
            GOTO      ON                 ;Repeat
END
```

How does it work?

- We check to see if the switch has been pressed (clear). If not GOTO ON. If it has skip that line and turn on the LED on B0. The code is:

```
ON        BTFSC        PORTA,0 ;Check button pressed.
          GOTO         ON
          BSF          PORTB,0 ;Turn on LED.
```

- Wait 5 seconds. The 5 second delay has been included for you in the subroutine section. Code:

```
CALL         DELAY5
```

- Turn the LED off and go back to the beginning. Code:

```
BCF       PORTB,0         ;Turn off LED.
GOTO      ON
```

Try this next problem for yourselves, before looking at the solution.

Problem 1: Using Port A bit 0 as a start button and outputs on Port B bits 0–3, switch on Port B bits 0 and 2 for ¼ second, switch off bits 0 and 2. Switch on Port B bits 1 and 3 for ¼ second, switch off bits 1 and 3. Repeat continuously. The ¼ second delay is provided for you.

The flowchart for the solution to this problem is shown in Figure 3.4.

Program solution to Problem 1 for the 16F84

```
;PROBLEM1.ASM

; EQUATES SECTION

TMR0      EQU     1        ;means TMR0 is file 1.
STATUS    EQU     3        ;means STATUS is file 3.
PORTA     EQU     5        ;means PORTA is file 5.
PORTB     EQU     6        ;means PORTB is file 6.
ZEROBIT   EQU     2        ;means ZEROBIT is bit 2.
COUNT     EQU     0CH      ;means COUNT is file 0C,
                           ;a register to count events.
;****************************************************

LIST      P=16F84  ;we are using the 16F84.
ORG       0        ;the start address in memory is 0
GOTO      START    ;goto start!
;****************************************************
```

;SUBROUTINE SECTION.

;0.25 second delay.

```
DELAY       CLRF        TMR0                    ;START TMR0.
LOOPA       MOVF        TMR0,W                  ;READ TMR0 INTO W.
            SUBLW       .8                      ;TIME – 8
            BTFSS       STATUS,ZEROBIT          ;Check TIME-W = 0
            GOTO        LOOPA                   ;Time is not = 8.
            RETLW       0                       ;Time is 8, return.
```

;***
;CONFIGURATION SECTION

```
START       BSF         STATUS,5                ;Turns to Bank1.
            MOVLW       B'00011111'             ;5 bits of PORTA are I/P
            TRIS        PORTA
            MOVLW       B'00000000'
            TRIS        PORTB                   ;PORTB is OUTPUT
            MOVLW       B'00000111'             ;Prescaler is /256
            OPTION                              ;TIMER is 1/32 secs.
            BCF         STATUS,5                ;Return to Bank0.
            CLRF        PORTA                   ;Clears PortA.
            CLRF        PORTB                   ;Clears PortB.
```

;***
;Program starts now.

```
ON          BTFSC       PORTA,0                 ;Check button pressed.
            GOTO        ON
REPEAT      MOVLW       B'00000101'
            MOVWF       PORTB                   ;Turn on bits 0 and 2
            CALL        DELAY                   ;¼ second delay
            MOVLW       B'00001010'
            MOVWF       PORTB                   ;Turn on bits 1 and 3
            CALL        DELAY                   ;¼ second delay
            GOTO        REPEAT                  ;Repeat
END
```

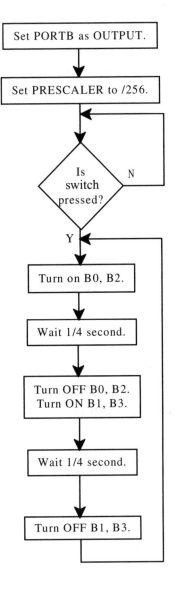

Figure 3.4 Flowchart for Problem 1

How does it work?

- Wait for the switch on PORTA,0 to clear, with BTFSC PORTA,0 then skip to
- MOVLW B'00000101' this sets up the data in the W register.
- MOVWF PORTB transfers the W register to PORTB and puts 5v on B0 and B2 only.
- CALL DELAY waits for ¼ second.
- MOVLW B'00001010' this sets up the data in the W register.
- MOVWF PORTB transfers the W register to PORTB and puts 5v on B1 and B3 only.
- CALL DELAY waits for ¼ second.
- GOTO REPEAT sends the program back to (my) label, REPEAT. This will keep the lights flashing all the time without checking the switch again.

Question: How do we make the program look at the switch, so that we can control whether or not the program repeats?

Answer: Instead of GOTO REPEAT use GOTO BEGIN. The program will then goto the label BEGIN instead of REPEAT and will wait for the switch to be Clear before repeating.

Extra work: Try and make the flashing routine more interesting by adding more combinations.

Scanning (using multiple inputs)

Scanning (also called polling) is when the microcontroller looks at the condition of a number of inputs in turn and executes a section of program depending on the state of those inputs. Applications include:

- Burglar alarms – when sensors are monitored and a siren sounds either immediately or after a delay depending on which input is active.
- Keypad scanning – a key press could cause an LED to light, a buzzer to sound or a missile to be launched. Just do not press the wrong key!

Let's consider a simple example.

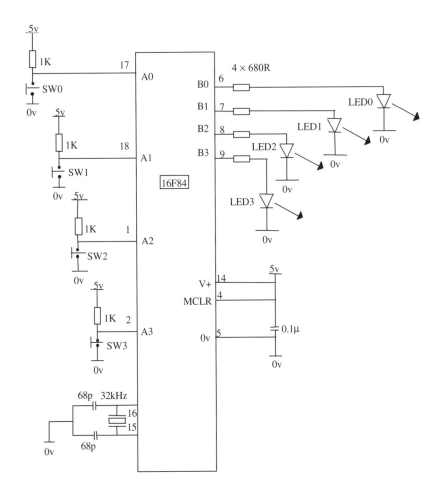

Figure 3.5 Switch scanning circuit

Switch scanning

Design a circuit so that if a switch is pressed a corresponding LED will light, i.e.

If SW0 is Hi (logic 1 or Set) then LED0 is on.
If SW0 is Low (logic 0 or Clear) then LED0 is off.
If SW1 is Hi (logic 1 or Set) then LED1 is on.
If SW1 is Low (logic 0 or Clear) then LED1 is off.
Etc.

The circuit diagram for this is shown in Figure 3.5 and the corresponding flow-chart in Figure 3.6.

The program solution for the switch scan

```
;SWSCAN.ASM

;EQUATES SECTION

TMR0        EQU      1              ;means TMR0 is file 1.
STATUS      EQU      3              ;means STATUS is file 3.
PORTA       EQU      5              ;means PORTA is file 5.
PORTB       EQU      6              ;means PORTB is file 6.
;***********************************************************
LIST        P=16F84                 ;we are using the 16F84.
ORG         0                       ;the start address in memory is 0
GOTO        START                   ;goto start!

;***********************************************************

;CONFIGURATION SECTION

START       BSF      STATUS,5        ;Turns to Bank1.
            MOVLW    B'00011111'     ;5bits of PORTA are I/P
            TRIS     PORTA
            MOVLW    B'00000000'
            TRIS     PORTB           ;PORTB is OUTPUT
            MOVLW    B'00000111'     ;Prescaler is /256
            OPTION                   ;TIMER is 1/32 secs.
            BCF      STATUS,5        ;Return to Bank0.
            CLRF     PORTA           ;Clears PortA.
            CLRF     PORTB           ;Clears PortB.

;***********************************************************
;Program starts now.

SW0                  BTFSC    PORTA,0    ;Switch0 pressed?
                     GOTO     TURNON0    ;Yes
                     BCF      PORTB,0    :No, Switch off LED0.

SW1                  BTFSC    PORTA,1    ;Switch1 pressed?
                     GOTO     TURNON1    ;Yes
                     BCF      PORTB,1    :NO Switch off LED1.
```

SW2	BTFSC	PORTA,2	;Switch2 pressed?
	GOTO	TURNON2	;Yes
	BCF	PORTB,2	:NO Switch off LED2.

SW3	BTFSC	PORTA,3	;Switch3 pressed?
	GOTO	TURNON3	;Yes
	BCF	PORTB,1	:NO Switch off LED3.

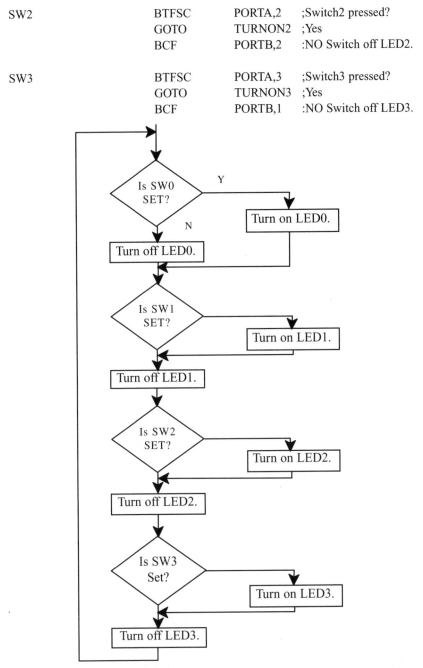

Figure 3.6 Flowchart for switch scan

```
            GOTO      SW0        ;Rescan.

TURNON0     BSF       PORTB,0    ;Turn on LED0
            GOTO      SW1

TURNON1     BSF       PORTB,1    ;Turn on LED1
            GOTO      SW2
TURNON2     BSF       PORTB,2    ;Turn on LED2
            GOTO      SW3

TURNON3     BSF       PORTB,3    ;Turn on LED3
            GOTO      SW0
END
```

How does it work?

- SW0 is checked first with the instruction BTFSC PORTA,0. If the switch is closed when the program is executing this line then we GOTO TURNON0. That is the program jumps to the label TURNON0 which turns on LED0 and then jumps the program back to check SW1 at, of course, the label SW1.
- SW1 is then checked in the same manner and then SW2 and SW3.

Suppose we press the switch when the program is not looking at it. The program lines are being executed at ¼ of the clock frequency, i.e. 32 768 Hz, that is 8192 lines a second. The program will always catch you!

Try modifying the program so that the switches can flash four different routines, e.g. SW0 flashes all lights on and off five times for 1 second.

INPUTS								OUTPUTS	
x	x	x	A4	A3	Room A2	Water A1	OverH A0	Heater B1	Fan B0
0	0	0	0	0	0	0	0	1	0
0	0	0	0	0	0	0	1	0	1
0	0	0	0	0	0	1	0	1	1
0	0	0	0	0	0	1	1	0	1
0	0	0	0	0	1	0	0	0	0
0	0	0	0	0	1	0	1	0	1
0	0	0	0	0	1	1	0	0	0
0	0	0	0	0	1	1	1	0	1

Figure 3.7 Truth table for the hot air system

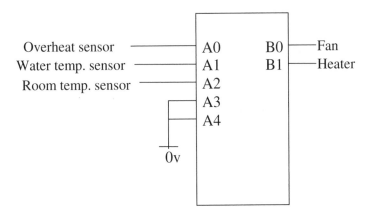

Figure 3.8 Block diagram for the hot air system

Control application – a hot air blower

The preceding section outlined how to monitor inputs by looking at them in turn. This application will 'read' all the bits on the port at once, because we will be concerned with particular combinations of inputs rather than individual ones. The bits on the input port will be 0s or 1s and we can treat this binary pattern like any other number in a file.

Consider a controller for a hot air radiator. When the water is warm the fan will blow the warm air into the room. The heater and fan are controlled by three temperature sensors: (a) a room temperature sensor, (b) a boiler water temperature sensor and (c) a safety overheating sensor. The truth table for the system is shown in Figure 3.7, where a 1 means hot and a 0 means cold for the sensors. The block diagram for the system is shown in Figure 3.8

Note: A3 and A4 are inputs and need to be connected to 0v. Do not leave them floating – you would not know if they were 0 or 1! Although they are not being used they are still being read.

There are eight input conditions from our three sensors. So all eight must be checked to determine which condition is true.

Consider the first condition A2 = A1 = A0 = 0. i.e. PORTA reads 0000 0000. How do we know that PORTA is 0000 0000? We do not have an instruction that says 'is PORTA equal to 0000 0000' or any other value for that matter. So we need to look at our 35 instructions and come up with a way of finding out what is the binary value of PORTA.

We check for this condition by subtracting 00000000 from it, if the answer is zero then PORTA reads 00000000, i.e. 0000 0000 – 0000 0000 = 0 (obviously). But how do we subtract the two numbers and how do we know if the answer is zero? This is a very important piece of programming so read the next few lines *carefully.*

- We first read PORTA into the W register with the instruction MOVF PORTA,W, this moves the data (settings of the switches, 1s or 0s) into W.
- We then subtract the number we are looking for, in this case 00000000, from W.
- We then need to know if the answer to this subtraction is zero. If it is then the value on PORTA was 00000000. If the answer is not zero then the value of the data on PORTA was not zero.
- So is the answer zero? Yes or no? The answer is held in a register called the status register, in bit 2 of this register, called the zerobit. If the zerobit, called a flag, is 1, it is indicating that the statement is true, the calculation was zero. If the zerobit is 0, this indicates the statement is false, the answer was not zero.
- We test the zerobit in the status register as we tested the bit on the switch connected to PORTA at the start of this chapter. We use the command BTFSC and the instruction BTFSC STATUS,ZEROBIT. If the zerobit is clear we skip the next instruction; if it is set we have a match and do not skip.

The code for this is:

```
MOVLW      B'00000000'        ;put 00000000 in W
SUBWF      PORTA              ;subtract W from PORTA
BTFSC      STATUS,ZEROBIT     ;PORTA=00000000?
CALL       CONDA              ;yes
```

CONDA is short for condition A where we require the heater on and the fan off.

- To check for A2 = A1 = 0 and A0 = 1 we subtract 00000001. To check for the next condition A2 = 0, A1 = 1, A0 = 0 we subtract 00000010, and so on for the other five conditions.

```
MOVLW      B'00000001'        ;put 00000001 in W
SUBWF      PORTA              ;subtract W from PORTA
BTFSS      STATUS,ZEROBIT     ;PORTA=00000001?
CALL       CONDB              ;yes
etc.
```

The complete opcode for the program CONTROL.ASM

```
;CONTROL.ASM
;*************************************************************
;
```

;EQUATES SECTION

TMR0	EQU	1	;means TMR0 is file 1.
STATUS	EQU	3	;means STATUS is file 3.
PORTA	EQU	5	;means PORTA is file 5.
PORTB	EQU	6	;means PORTB is file 6.
ZEROBIT	EQU	2	;means ZEROBIT is bit 2.

;**

LIST	P=16F84	;we are using the 16F84.
ORG	0	;the start address in memory is 0
GOTO	START	;goto start!

;**

;SUBROUTINE SECTION.

CONDA	BCF	PORTB,0	;turns fan off
	BSF	PORTB,1	;turns heater on
	RETLW	0	
CONDB	BSF	PORTB,0	;turns fan on
	BCF	PORTB,1	;turns heater off
	RETLW	0	
CONDC	BSF	PORTB,0	;turns fan on
	BSF	PORTB,1	;turns heater on
	RETLW	0	
CONDD	BCF	PORTB,0	;turns fan off
	BCF	PORTB,1	;turns heater off
	RETLW	0	

;**

;CONFIGURATION SECTION

START	BSF	STATUS,5	;Turns to Bank1.
	MOVLW	B'00011111'	;5 bits of PORTA are I/P
	TRIS	PORTA	
	MOVLW	B'00000000'	
	TRIS	PORTB	;PORTB is OUTPUT
	BCF	STATUS,5	;Return to Bank0.
	CLRF	PORTA	;Clears PortA.
	CLRF	PORTB	;Clears PortB.

```
;************************************************************
;Program starts now.

BEGIN     MOVLW      B'00000000'        ;put 00000000 in W
          SUBWF      PORTA              ;PORTA – W
          BTFSC      STATUS,ZEROBIT     ;PORTA=00000000?
          CALL       CONDA              ;yes

          MOVLW      B'00000001'        ;put 00000001 in W
          SUBWF      PORTA              ;PORTA – W
          BTFSC      STATUS,ZEROBIT     ;PORTA=00000001?
          CALL       CONDB              ;yes

          MOVLW      B'00000010'        ;put 00000010 in W
          SUBWF      PORTA              ;PORTA – W
          BTFSC      STATUS,ZEROBIT     ;PORTA=00000010?
          CALL       CONDC              ;yes

          MOVLW      B'00000011'        ;put 00000011 in W
          SUBWF      PORTA              ;PORTA – W
          BTFSC      STATUS,ZEROBIT     ;PORTA=00000011?
          CALL       CONDB              ;yes

          MOVLW      B'00000100'        ;put 00000100 in W
          SUBWF      PORTA              ;PORTA – W
          BTFSC      STATUS,ZEROBIT     ;PORTA=00000100?
          CALL       CONDD              ;yes

          MOVLW      B'00000101'        ;put 00000101 in W
          SUBWF      PORTA              ;PORTA – W
          BTFSC      STATUS,ZEROBIT     ;PORTA=00000101?
          CALL       CONDB              ;yes

          MOVLW      B'00000110'        ;put 00000110 in W
          SUBWF      PORTA              ;PORTA – W
          BTFSC      STATUS,ZEROBIT     ;PORTA=00000110?
          CALL       CONDD              ;yes

          MOVLW      B'00000111'        ;put 00000111 in W
          SUBWF      PORTA              ;PORTA – W
          BTFSC      STATUS,ZEROBIT     ;PORTA=00000111?
          CALL       CONDB              ;yes

          GOTO       BEGIN
```

END

The program can be checked by using switches for the input sensors and LEDs for the outputs.

There is more than one way of skinning a cat – another way of writing this program is shown later in the section on look up tables, Chapter 5.

Analysing HEADER84.ASM

Now that we have looked at a number of applications we are ready to understand HEADER84.ASM listed in full in Chapter 2.

• The header starts with a title that includes the name of the file, this is useful when you are printing it out; it details what the program is doing.
  ; HEADER84.ASM for 16F84.This sets PORTA as an INPUT (NB 1
  ;   means input) and PORTB as an OUTPUT
  ;   (NB 0 means output). The OPTION
  ;   register is set to /256 to give timing pulses
  ;   of 1/32 of a second.
  ;   1second and 0.5 second delays are
  ;   included in the subroutine section.

  ;***
  ;
• The Equates Section tells the software what numbers your words represent. When you write your program you use mnemonics such as PORTA, PORTB, TMR0, STATUS, ZEROBIT, COUNT, MYAGE. The Assembler Program does not understand your words; it is looking for the file number or the bit number. You have to tell it what these mean in the Equates Section, i.e. COUNT is file 0C, PortA is file 5, the STATUS register is file 3, ZEROBIT is bit 2, etc. The memory map of the 16F84 in Appendix D shows the addresses of the registers and user files. The file with address 0C is the first of the user files and I have called it COUNT, it stores the number of times certain events have happened in my program. I could have file 0D as COUNT2, file 0E as COUNT3, file 0F as SECONDS or WAIT, etc.

;EQUATES SECTION

TMR0	EQU	1	;means TMR0 is file 1.
STATUS	EQU	3	;means STATUS is file 3.
PORTA	EQU	5	;means PORTA is file 5.
PORTB	EQU	6	;means PORTB is file 6.
ZEROBIT	EQU	2	;means ZEROBIT is bit 2.
COUNT	EQU	0CH	;means COUNT is file 0C,

;a register to count events.

- What chip are we using?

```
LIST      P=16F84      ;we are using the 16F84.
ORG       0            ;the start address in memory is 0
GOTO      START        ;goto start!
```

LIST P = 16F84 tells the assembler what chip to assemble the code for.

ORG 0 means put the next line of code into EPROM address 0, then follow with next line in address 1, etc.

GOTO START makes the program bypass the subroutine section and GOTO the label START which is where the device is configured before executing the body of the program. The instruction GOTO START is placed in EPROM address 0 by ORG 0.

The line DELAY1 CLRF TMR0 is then placed in EPROM address 1, etc.

- Subroutine Section.

The Subroutine Section consists of two subroutines DELAY1 and DELAYP5. A subroutine is a section of program, which is used a number of times instead of rewriting it and using up program memory. Just call it e.g., CALL DELAY1, at the end you RETURN to the program in the position you left it. The stack is the register that remembers where you came from and takes you back. The DELAY1 code is:

```
DELAY1    CLRF     TMR0                   ;Start TMR0.
LOOPA     MOVF     TMR0,W                 ;Read TMR0 into W.
          SUBLW    .32                    ;TIME – 32
          BTFSS    STATUS,ZEROBIT         ;Check TIME-W = 0
          GOTO     LOOPA                  ;Time is not = 32.
          RETLW    0                      ;Time is 32, return.
```

DELAY1 starts by clearing the register TMR0 (timer 0), with CLRF TMR0, i.e. CleaR the File TMR0. This sets the timer to zero and will be counting TMR0 pulses every 1/32 of a second.

LOOPA MOVF TMR0,W is move file TMR0 into the Working register, W.

We want to know when TMR0 is 32 because then we will have had 32 TIMER0 pulses, which is 1 second. This is done with a subtraction as in the example earlier in this chapter, in the section on the hot air blower.

The label LOOPA is there because we keep returning to it until TMR0 reaches the required value.

There is no instruction, that asks the micro is TMR0 equal to 32? So we have to use the instructions available. We subtract a number from W and ask is the answer 0. If, for example, we subtract 135; from W and the answer is 0 then

W contained 135; if the answer was not 0 then W did not contain 135. The status register contains a bit called a zerobit, it is bit 2. Notice in the EQUATES section I have put ZEROBIT EQU 2, so I can use ZEROBIT in my code instead of 2 – I would soon forget what the 2 was supposed to mean. The zerobit is set to a 1 when the result of a previous calculation is 0. So a 1 means result was 0! Think of this as a flag (because that's what it is called), the flag is waving (a 1) to indicate the result is zero. We can test this zerobit, i.e. look at it and see if it is a 1 or 0. We can skip the next instruction if it is set (a zero has occurred) by BTFSS STATUS,ZEROBIT or skip if clear (a zero has not occurred) by BTFSC STATUS,ZEROBIT. Doesn't this read better than BTFSC 3,2. STATUS is Register3, ZEROBIT is bit 2?
Let's look at this subroutine again.

DELAY1	CLRF	TMR0	;START TMR0.
LOOPA	MOVF	TMR0,W	;READ TMR0 INTO W.
	SUBLW	.32	;TIME – 32
	BTFSS	STATUS,ZEROBIT	;Check TIME – W = 0
	GOTO	LOOPA	;Time is not = 32.
	RETLW	0	;Time is 32, return.

- We clear TMR0 (CLRF TMR0).
- Then move TMR0 into W (MOVF TMR0,W).
SUBTRACT 32 from W which now holds TMR0 value (SUBLW 32).
- If W (hence TMR0) is 32 the zerobit is set, we skip the next instruction and return from the subroutine with 0 in W (RETLW 0).
- If W is not 32 then we do not skip and we GOTO LOOPA and put TMR0 in W and repeat until TMR0 is 32.

DELAYP5 is a similar code but TMR0 now is only allowed to count up to 16, i.e. a half-second (with 32 pulses a second). Note: If you copy and paste to change the name of the subroutine from DELAY1 to DELAYP5, change the 32 to 16 and do not forget to change LOOPA to LOOPB. You cannot goto room 27 if there are two room 27s!

- Configuration Section.

START	BSF	STATUS,5	;Turns to Bank1.
	MOVLW	B'00011111'	;5 bits of PORTA are I/P
	TRIS	PORTA	
	MOVLW	B'00000000'	
	TRIS	PORTB	;PORTB is OUTPUT
	MOVLW	B'00000111'	;Prescaler is /256
	OPTION		;TIMER is 1/32 sec.

```
BCF       STATUS,5      ;Return to Bank0.
CLRF      PORTA         ;Clears PortA.
CLRF      PORTB         ;Clears PortB.
```

The instruction BSF STATUS,5 sets bit 5 in the status register. As you can see from the explanation of the status register bits in Appendix D, bit 5 is a page select bit which selects page 1 giving us access to the registers in the page 1 column of the memory map in Appendix D. The reason for pages is that we have an 8-bit micro. 8 bits can only address 255 files so to identify a file we have it on a page, like a line in a book, i.e. line 17 on page 40 instead of line 2475.

```
MOVLW     B'00011111'   ;5bits of PORTA are I/P
TRIS      PORTA
```

These two lines move 11111 into the data direction register to set the 5 bits of PORTA as inputs. The 11111 is first moved to W (MOVLW B'00011111') and then into the data direction register with TRIS PORTA. A 1 signifies an input, a 0 an output.

```
MOVLW     B'00000000'   ;8bits of PORTB are O/P
TRIS      PORTB
```

These two lines move 00000000 into the data direction register to set the 8 bits of PORTB as outputs. The 000000 is first moved to W and then into the data direction register with TRIS PORTB.

PortA and PortB can be configured differently if required. For example, to make the lower 4 bits of PortB outputs and the upper 4 bits inputs – alter the two lines of the program with:

```
MOVLW     B'11110000'
TRIS      PORTB
```

The header also sets the internal clock to divide by 256, i.e. a 32.768kHz clock gives a program execution of 32.768kHz/4 = 8.192kHz. If the prescaler is set to divide by 256 this gives timing pulses of 32 a second.
The prescaler is configured with the two lines:

```
MOVLW     B'00000111'   ;Prescaler is /256
OPTION                  ;TIMER is 1/32 sec.
```

The OPTION register can be altered in the header to give faster timing pulses if required, as described in the OPTION register section in Appendix D. The line

4
Keypad scanning

There are no new instructions used in this chapter.

Keypads are an excellent way of entering data into the microcontroller. The keys are usually numbered but they could be labelled as function keys, for example in a remote control handset in a TV to adjust the sound or colour, etc.

As well as remote controls, keypads find applications in burglar alarms, door entry systems, calculators, microwave ovens, etc. So there are no shortage of applications for this section.

Keypads are usually arranged in a matrix format to reduce the number of I/O connections.

A 12 key keypad is arranged in a 3×4 format requiring seven connections. A 16 key keypad is arranged in a 4×4 format requiring eight connections.

12 key keypad

Consider the 12 key keypad. This is arranged in three columns and four rows as shown in Figure 4.1. There are seven connections to the keypad – C1, C2, C3, R1, R2, R3 and R4.

This connection to the micro is shown in Figure 4.2.

The keypad works in the following way:

If, for example, key 6 is pressed then B2 will be joined to B4. For key 1 B0 would be joined to B3, etc. as shown in Figure 4.2.

	Column1, C1	Column2, C2	Column3, C3
Row1, R1	1	2	3
Row2, R2	4	5	6
Row3, R3	7	8	9
Row4, R4	*	0	#

Figure 4.1 12 key keypad

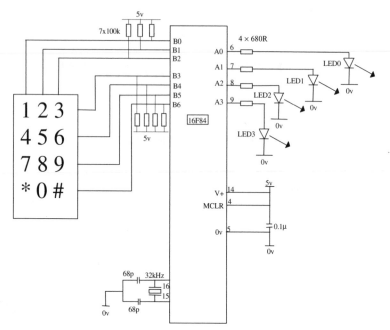

Figure 4.2 Keypad connection to the microcontroller

The micro would set B0 low and scan B3, B4, B5 and B6 for a low to see if keys 1, 4, 7 or * had been pressed.

The micro would then set B1 low and scan B3, B4, B5 and B6 for a low to see if keys 2, 5, 8 or 0 had been pressed.

Finally B2 would be set low and B3, B4, B5 and B6 scanned for a low to see if keys 3, 6, 9 or # had been pressed.

Programming example for the keypad

As a programming example, when key 1 is pressed a binary 1 is displayed on PORTA, when key 2 is pressed a binary 2 is displayed on PORTA, etc.
Key 0 displays 10. Key * displays 11. Key # displays 12.

This program could be used as a training aid for decimal to binary conversion. The flowchart is shown in Figure 4.3.

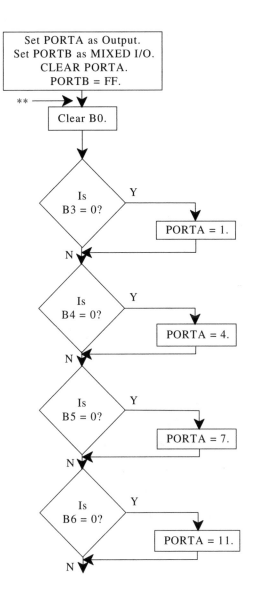

Figure 4.3 Keypad scanning flowchart

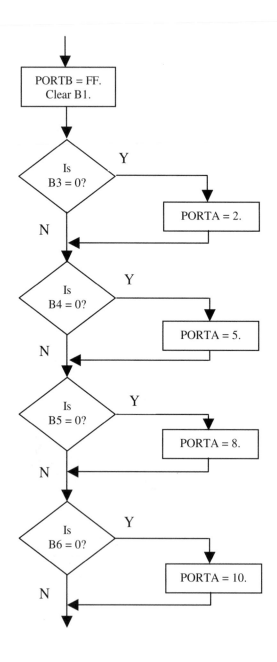

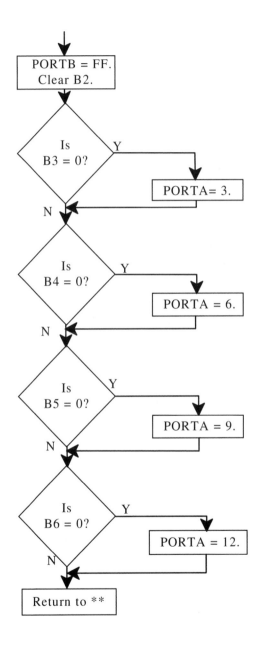

The program listing for the keypad example

;KEYPAD.ASM

;EQUATES SECTION

```
STATUS      EQU      3                    ;means STATUS is file 3.
PORTA       EQU      5                    ;means PORTA is file 5.
PORTB       EQU      6                    ;means PORTB is file 6.
```

;**
;

```
LIST        P=16F84  ;we are using the 16F84.
ORG         0        ;the start address in memory is 0
GOTO        START    ;goto start!
```

;**
;

;CONFIGURATION SECTION

```
START       BSF      STATUS,5             ;Turns to Bank1.
            MOVLW    B'00000000'          ;PORTA is OUTPUT
            TRIS     PORTA
            MOVLW    B'11111000'
            TRIS     PORTB                ;PORTB is mixed I/O.
            BCF      STATUS,5             ;Return to Bank0.
            CLRF     PORTA                ;Clears PortA.
            CLRF     PORTB                ;Clears PortB.
```

;**
;
;Program starts now.

```
COLUMN1     BCF      PORTB,0              ;Clear B0
            BSF      PORTB,1              ;Set B1
            BSF      PORTB,2              ;Set B2

CHECK1      BTFSC    PORTB,3              ;Is B3 Clear?
            GOTO     CHECK4               ;No
            MOVLW    .1                   ;Yes, output 1.
            MOVWF    PORTA
CHECK4      BTFSC    PORTB,4              ;Is B4 Clear?
            GOTO     CHECK7               ;No
            MOVLW    .4                   ;Yes, output 4.
```

	MOVWF	PORTA	
CHECK7	BTFSC	PORTB,5	;Is B5 Clear?
	GOTO	CHECK11	;No
	MOVLW	.7	;Yes, output 7.
	MOVWF	PORTA	
CHECK11	BTFSC	PORTB,6	;Is B6 Clear?
	GOTO	COLUMN2	;No
	MOVLW	.11	;Yes, output 11.
	MOVWF	PORTA	

	BSF	PORTB,0	
COLUMN2	BSF	PORTB,0	;Set B0
	BCF	PORTB,1	;Clear B1
	BSF	PORTB,2	;Set B2
CHECK2	BTFSC	PORTB,3	;Is B3 Clear?
	GOTO	CHECK5	;No
	MOVLW	.2	;Yes, output 2.
	MOVWF	PORTA	
CHECK5	BTFSC	PORTB,4	;Is B4 Clear?
	GOTO	CHECK8	;No
	MOVLW	.5	;Yes, output 5.
	MOVWF	PORTA	
CHECK8	BTFSC	PORTB,5	;Is B5 Clear?
	GOTO	CHECK10	;No
	MOVLW	.8	;Yes, output 8.
	MOVWF	PORTA	
CHECK10	BTFSC	PORTB,6	;Is B6 Clear?
	GOTO	COLUMN3	;No
	MOVLW	.10	;Yes, output 10.
	MOVWF	PORTA	

	BSF	PORTB,0	
COLUMN3	BSF	PORTB,0	;Set B0
	BSF	PORTB,1	;Set B1
	BCF	PORTB,2	;Clear B2
CHECK3	BTFSC	PORTB,3	;Is B3 Clear?
	GOTO	CHECK6	;No
	MOVLW	.3	;Yes, output 3.
	MOVWF	PORTA	
CHECK6	BTFSC	PORTB,4	;Is B4 Clear?
	GOTO	CHECK9	;No
	MOVLW	.6	;Yes, output 6.
	MOVWF	PORTA	
CHECK9	BTFSC	PORTB,5	;Is B5 Clear?
	GOTO	CHECK12	;No
	MOVLW	.9	;Yes, output 9.

```
              MOVWF    PORTA
CHECK12       BTFSC    PORTB,6        ;Is B6 Clear?
              GOTO     COLUMN1        ;No
              MOVLW    .12            ;Yes, output 12.
              MOVWF    PORTA
              GOTO     COLUMN1        ;Start scanning again.
```

END.

How does the program work?

Port configuration: The first thing to note about the keypad circuit is that the PORTA pins are being used as outputs. On PORTB, pins B0, B1 and B2 are outputs and B3, B4, B5 and B6 are inputs. So PORTB is a mixture of inputs and outputs. The HEADER84.ASM program has to be modified to change to this new configuration.

To change PORTA to an output port, the following two lines are used in the Configuration Section:

```
        MOVLW    B'00000000'       ;PORTA is OUTPUT
        TRIS     PORTA
```

To configure PORTB as a mixed input and output port the following two lines are used in the Configuration Section:

```
        MOVLW    B'11111000'
        TRIS     PORTB      ;PORTB is mixed I/O. B0,B1,B2 are O/P.
```

Scanning routine

The scanning routine looks at each individual key in turn to see if one is being pressed. Because it can do this so quickly it will notice we have pressed a key even if we press it quickly.

The scanning routine first of all looks at the keys in column1, i.e. 1, 4, 7 and *. It does this by setting B0 low, B1 and B2 high. If a 1 is pressed the B3 will be low, if a 1 is not pressed then B3 will be high. Because pressing a 1 connects B0 and B3. Similarly if 4 is pressed B4 will be low, if not B4 will be high. If 7 is pressed B5 will be low, if not B5 will be high. If * is pressed B6 will be low, if not B6 will be high.

In other words when we set B0 low if any of the keys in column1 are pressed then the corresponding input to the microcontroller will go low and the program will output the binary number equivalent of the key that has been pressed.

If none of the keys in column1 are pressed then we move onto column2.

The code for scanning column1 is as follows – these three lines set up PORTB with B0 = 0, B1 = 1 and B2 = 1:

```
COLUMN1    BCF     PORTB,0     ;Clear B0
           BSF     PORTB,1     ;Set B1
           BSF     PORTB,2     ;Set B2
```

These next four lines test input B3 to see if it is clear; if it is then a 1 is placed on PORTA, and the program continues. If B3 is set then we proceed to Check to see if key 4 has been pressed, with CHECK4.

```
CHECK1     BTFSC   PORTB,3     ;Is B3 Clear?
           GOTO    CHECK4      ;No
           MOVLW   .1          ;Yes, output 1
           MOVWF   PORTA       ;to PORTA
```

These next four lines test input B4 to see if it is clear; if it is then a 4 is placed on PORTA, and the program continues. If B4 is set then we proceed to Check to see if key 7 has been pressed, with CHECK7.

```
CHECK4     BTFSC   PORTB,4     ;Is B4 Clear?
           GOTO    CHECK7      ;No
           MOVLW   .4          ;Yes, output 4.
           MOVWF   PORTA
```

These next four lines test input B5 to see if it is clear; if it is then a 7 is placed on PORTA, and the program continues. If B5 is set then we proceed to Check to see if key * has been pressed, with CHECK11.

```
CHECK7     BTFSC   PORTB,5     ;Is B5 Clear?
           GOTO    CHECK11     ;No
           MOVLW   .7          ;Yes, output 7.
           MOVWF   PORTA
```

These next four lines test input B6 to see if it is clear; if it is then an 11 is placed on PORTA, and the program continues. If B5 is set then we proceed to Check the keys in column2, with COLUMN2.

```
CHECK11    BTFSC   PORTB,6     ;Is B6 Clear?
           GOTO    COLUMN2     ;No
```

```
MOVLW    .11                    ;Yes, output 11.
MOVWF    PORTA
```

These three lines set up PORTB with B0 = 1, B1 = 0 and B2 = 1.

```
COLUMN2    BSF    PORTB,0       ;Set B0
           BCF    PORTB,1       ;Clear B1
           BSF    PORTB,2       ;Set B2
```

We then check to see if key 2 has been pressed by testing to see if B3 is clear; if it is then a 2 is placed on PORTA and the program continues. If B3 is set then we proceed with CHECK5. This code is:

```
CHECK2     BTFSC   PORTB,3      ;Is B3 Clear?
           GOTO    CHECK5       ;No
           MOVLW   .2           ;Yes, output 2.
           MOVWF   PORTA
```

The program continues in the same manner checking 5, 8 and 10 (0), then moving onto column3 to check for 3, 6, 9 and 12 (#). After completing the scan the program then goes back to continue the scan again.

It takes about 45 lines of code to complete a scan of the keypad. With a 32 768 Hz crystal the lines of code are executed at ¼ of this speed, i.e. 8192 lines per second. So the scan time is 45/8192 = 0.0055 second. This is why no matter how quickly you press the key the microcontroller will be able to detect it.

Security code

Probably one of the most useful applications of a keypad is to enter a code to turn something on and off such as a burglar alarm or door entry system.

In the following program KEYS3.ASM the subroutine SCAN scans the keypad, waits for a key to be pressed, waits 0.1 seconds for the bouncing to stop, waits for the key to be released, waits 0.1 seconds for the bouncing to stop and then returns with the key number in W which can then be transferred into a file.

This is then used as a security code to turn on an LED (PORTA,0) when three digits (137) have been pressed and turn the LED off again when the same three digits are pressed. You can of course use any three digits.

```
;KEYS3.ASM
;EQUATES SECTION

ZEROBIT      EQU 2
TMR0         EQU 1
STATUS       EQU 3          ;means STATUS is file 3.
PORTA        EQU 5          ;means PORTA is file 5.
PORTB        EQU 6          ;means PORTB is file 6.
NUM1         EQU 0CH
NUM2         EQU 0DH
NUM3         EQU 0EH

;**********************************************************

LIST P=16F84                ;we are using the 16F84.
ORG 0                       ;the start address in memory is 0
GOTO START                  ;goto start!

;**********************************************************
;SUBROUTINE SECTION

SCAN         NOP

COLUMN1      BCF      PORTB,0        ;Clear B0
             BSF      PORTB,1        ;Set B1
             BSF      PORTB,2        ;Set B2

CHECK1       BTFSC    PORTB,3        ;Is B3 Clear?
             GOTO     CHECK4         ;No
             CALL     DELAYP1
CHECK1A      BTFSS    PORTB,3
             GOTO     CHECK1A
             CALL     DELAYP1
             RETLW    .1

CHECK4       BTFSC    PORTB,4        ;Is B4 Clear?
             GOTO     CHECK7         ;No
             CALL     DELAYP1
CHECK4A      BTFSS    PORTB,4
             GOTO     CHECK4A
             CALL     DELAYP1
             RETLW    .4

CHECK7       BTFSC    PORTB,5        ;Is B5 Clear?
```

	GOTO	CHECK11	;No
	CALL	DELAYP1	
CHECK7A	BTFSS	PORTB,5	
	GOTO	CHECK7A	
	CALL	DELAYP1	
	RETLW	.7	

CHECK11	BTFSC	PORTB,6	;Is B6 Clear?
	GOTO	COLUMN2	;No
	CALL	DELAYP1	
CHECK11A	BTFSS	PORTB,6	
	GOTO	CHECK11A	
	CALL	DELAYP1	
	RETLW	.11	

COLUMN2	BSF	PORTB,0	;Set B0
	BCF	PORTB,1	;Clear B1
	BSF	PORTB,2	;Set B2

CHECK2	BTFSC	PORTB,3	;Is B3 Clear?
	GOTO	CHECK5	;No
	CALL	DELAYP1	
CHECK2A	BTFSS	PORTB,3	
	GOTO	CHECK2A	
	CALL	DELAYP1	
	RETLW	.2	;Yes, output 2.

CHECK5	BTFSC	PORTB,4	;Is B4 Clear?
	GOTO	CHECK8	;No
	CALL	DELAYP1	
CHECK5A	BTFSS	PORTB,4	
	GOTO	CHECK5A	
	CALL	DELAYP1	
	RETLW	.5	;Yes, output 5.

CHECK8	BTFSC	PORTB,5	;Is B5 Clear?
	GOTO	CHECK0	;No
	CALL	DELAYP1	
CHECK8A	BTFSS	PORTB,5	
	GOTO	CHECK8A	
	CALL	DELAYP1	
	RETLW	.8	;Yes, output 8.

| CHECK0 | BTFSC | PORTB,6 | ;Is B6 Clear? |

	GOTO	COLUMN3	;No
	CALL	DELAYP1	
CHECK0A	BTFSS	PORTB,6	
	GOTO	CHECK0A	
	CALL	DELAYP1	
	RETLW	0	;Yes, output 10.

COLUMN3	BSF	PORTB,0	;Set B0
	BSF	PORTB,1	;Set B1
	BCF	PORTB,2	;Clear B2

CHECK3	BTFSC	PORTB,3	;Is B3 Clear?
	GOTO	CHECK6	;No
	CALL	DELAYP1	
CHECK3A	BTFSS	PORTB,3	
	GOTO	CHECK3A	
	CALL	DELAYP1	
	RETLW	.3	;Yes, output 3.

CHECK6	BTFSC	PORTB,4	;Is B4 Clear?
	GOTO	CHECK9	;No
	CALL	DELAYP1	
CHECK6A	BTFSS	PORTB,4	
	GOTO	CHECK6A	
	CALL	DELAYP1	
	RETLW	.6	;Yes, output 6.

CHECK9	BTFSC	PORTB,5	;Is B5 Clear?
	GOTO	CHECK12	;No
	CALL	DELAYP1	
CHECK9A	BTFSS	PORTB,5	
	GOTO	CHECK9A	
	CALL	DELAYP1	
	RETLW	.9	;Yes, output 9.

CHECK12	BTFSC	PORTB,6	;Is B6 Clear?
	GOTO	COLUMN1	;No
	CALL	DELAYP1	
CHECK12A	BTFSS	PORTB,6	
	GOTO	CHECK12A	
	CALL	DELAYP1	
	RETLW	.12	;Yes, output 12.

;3/32 second delay.

```
DELAYP1    CLRF      TMR0              ;Start TMR0.
LOOPD      MOVF      TMR0,W            ;Read TMR0 into W.
           SUBLW     .3                ;TIME - 3
           BTFSS STATUS,ZEROBIT        ;Check TIME-W = 0
           GOTO      LOOPD             ;Time is not = 3.
           RETLW     0                 ;Time is 3, return.
```

;***
;CONFIGURATION SECTION

```
START      BSF       STATUS,5          ;Turns to Bank1.
           MOVLW     B'00000000'       ;PORTA is OUTPUT
           TRIS      PORTA
           MOVLW     B'11111000'
           TRIS      PORTB             ;PORTB is mixed I/O.
           MOVLW     B'00000111'
           OPTION
           BCF       STATUS,5          ;Return to Bank0.
           CLRF      PORTA             ;Clears PortA.
           CLRF      PORTB             ;Clears PortB.
```

;***
;Program starts now.
;Enter 3 digit code here
```
           MOVLW     1                 ;First digit
           MOVWF     NUM1
           MOVLW     3                 ;Second digit
           MOVWF     NUM2
           MOVLW     7                 ;Third digit
           MOVWF     NUM3

BEGIN      CALL      SCAN              ;Get 1st number
           SUBWF     NUM1,W
           BTFSS     STATUS,ZEROBIT    ;IS NUMBER=1?
           GOTO      BEGIN             ;No

           CALL      SCAN              ;Get 2nd number
           SUBWF     NUM2,W
           BTFSS     STATUS,ZEROBIT    ;IS NUMBER=3?
           GOTO      BEGIN             ;No

           CALL      SCAN              ;Get 3rd number.
           SUBWF     NUM3,W
           BTFSS     STATUS,ZEROBIT    ;IS NUMBER=7?
```

```
           GOTO        BEGIN              ;No

           BSF         PORTA,0            ;Turn on LED, 137 entered

TURN_OFF   CALL        SCAN               ;Get 1st number again
           SUBWF       NUM1,W
           BTFSS       STATUS,ZEROBIT     ;IS NUMBER=1?
           GOTO        TURN_OFF           ;No

           CALL        SCAN               ;Get 2nd number
           SUBWF       NUM2,W
           BTFSS       STATUS,ZEROBIT     ;IS NUMBER=3?
           GOTO        TURN_OFF           ;No

           CALL        SCAN               ;Get 3rd number.
           SUBWF       NUM3,W
           BTFSS       STATUS,ZEROBIT     ;IS NUMBER=7?
           GOTO        TURN_OFF           ;No

           BCF         PORTA,0            ;Turn off LED.
           GOTO        BEGIN

END
```

How does the program work?

The ports are configured as in the previous code KEYPAD.ASM.

The KEYS3.ASM program looks for the first key press and then it compares the number pressed with the required number stored in a user file called NUM1. It then looks for the second key to be pressed. But because the microcontroller is so quick, the first number could be stored and the program looks for the second number, but our finger is still pressing the first number.

Antibounce routine

Also when a mechanical key is pressed or released it does not make or break cleanly, it bounces around. If the micro is allowed to, it is fast enough to see these bounces as key presses so we must slow it down.

- We look first of all for the switch to be pressed.
- Then wait 0.1 seconds for the switch to stop bouncing.
- We then wait for the switch to be released.
- We then wait 0.1 seconds for the bouncing to stop before continuing.

The switch has then been pressed and released indicating one action.

The 0.1 second delay is written in the header as DELAYP1.

Scan routine

The scan routine used in KEYS3.ASM is written into the subroutine.
When called it waits for a key to be pressed and then returns with the number
just pressed in W. It can be copied and used as a subroutine in any program
using a keypad.

- The scan routine checks for key presses as in the previous example KEY-PAD.ASM, column1 checks for the numbers 1, 4, 7 and 11 being pressed in turn.
- If the 1 is not pressed then the routine goes on to check for a 4.
- If the 1 is pressed then the routine waits 0.1 second for the bouncing to stop.
- The program then waits for the key to be released.
- Waits again 0.1 seconds for the bouncing to stop.
- Then returns with a value of 1 in W.

Code for CHECK1:

```
CHECK1    BTFSC    PORTB,3      ;Is B3 Clear? Pressed?
          GOTO     CHECK4       ;No
          CALL     DELAYP1      ;Antibounce delay, B3 clear
CHECK1A   BTFSS    PORTB,3      ;Is B3 Set? Released?
          GOTO     CHECK1A      ;No
          CALL     DELAYP1      ;Antibounce delay, B3 Set
          RETLW    .1           ;Return with 1 in W.
```

If numbers 4, 7 or 11 are pressed the routine will return with the corresponding
value in W. If no numbers in column1 are pressed then the scan routine continues on to column2 and column3. If no keys are pressed then the routine loops
back to the start of the scan routine to continue checking.

Storing the code

The code, i.e. 137, is stored in the files NUM1, NUM2, NUM3 with the following code:

```
          MOVLW    1            ;First digit
          MOVWF    NUM1
          MOVLW    3            ;Second digit
          MOVWF    NUM2
          MOVLW    7            ;Third digit
          MOVWF    NUM3
```

Checking for the correct code

- We first CALL SCAN to collect the first digit, which returns with the number pressed in W.
- We then Subtract the value of W from the first digit of our code stored in NUM1 with:

SUBWF NUM1,W.

This means SUBtract W from the File NUM1. The (,W) stores the result of the subtraction in W. Without (,W) the result would have been stored in NUM1 and the value changed.
- We then check to see if NUM1 and W are equal, i.e. a correct match. In this case the zerobit in the status register would be set. Indicating the result NUM1 − W = zero. This is done with:

BTFSS STATUS,ZEROBIT

We skip and carry on if it is set, i.e. a match. If it isn't we return to BEGIN to scan again.
- With a correct first press we then carry on checking for a second and if correct a third press to match the correct code.
- When the correct code is pressed we turn on our LED with:

BSF PORTA,0

- We then run through a similar sequence and wait for the code to turn off the LED.

Notice that if you enter an incorrect digit you return to BEGIN or TURN_OFF. If you forget what key you have pressed then press an incorrect one and start again. This program can be modified by adding a fourth digit to the program, then turn on the LED. In which case you use another user file called NUM4. You could of course use a different code for switching off the output.

You can also beep a buzzer for half a second to give yourself an audible feedback that you had pressed a button.

As an extra security measure you could wait for a couple of seconds if an incorrect key had been pressed, or wait for 2 minutes if three wrong numbers had been entered.

The keypad routine opens up many different circuit applications.

The SCAN routine can be copied and then pasted into any program using the keypad. Then when you CALL SCAN the program will return with the number pressed in W for you to do with it as you wish.

5
Program examples

New instructions used in this chapter:

- INCF
- INCFSZ
- DECF
- ADDWF

Counting events

Counting of course is a useful feature for any control circuit. We may wish to count the number of times a door has opened or closed, or count the number of pulses from a rotating disc. If we count cars into a car park we would increment a file count every time a car entered, using the instruction INCF COUNT. If we needed to know how many cars were in the car park we would have to reduce the count by one every time a car left. We would do this by DECF COUNT. To clear the user file COUNT to start we would CLRF COUNT. In this way the file COUNT would store the number of cars in the car park. If you prefer COUNT could be called CARS. It is a user file so you can call it what you like. Let's look at an application.

Design a circuit that will count 10 presses of a switch, then turn an LED on and reset when the next 10 presses are started. The hardware is that shown in Figure 3.1 with A0 as the switch input and B0 as the output to the LED.
There are two ways to count, UP and DOWN. We usually count up and know automatically when we have reached 10. A computer, however, knows when it reaches a count of 10 by subtracting the count from 10. If the answer is zero, then bingo. A simpler way, however, is to start at 10 and count down to zero – after 10 events we will have reached zero without doing a subtraction. Zero for the microcontroller is a really useful number.

The initial flowchart for this problem is shown in Figure 5.1.

To ensure that the LED is OFF after the switch is pressed for the 11th time put in TURN OFF LED after the switch is pressed, as shown in Figure 5.2.

Note: The switch will bounce and the micro is fast enough to count these

Counting flowchart

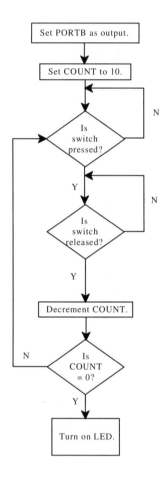

Figure 5.1 Initial counting flowchart

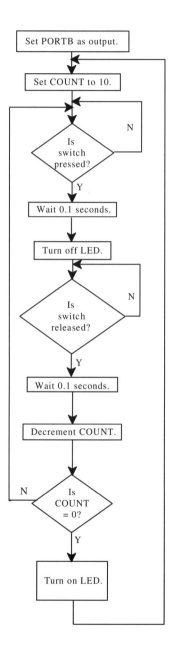

Figure 5.2 Final counting flowchart

bounces, thinking that the switch has been pressed several times. A 0.1 second delay is inserted after each switch operation to allow time for the bounces to stop. The final flowchart is shown in Figure 5.2.

The program for the counting circuit

;COUNT84.ASM

;EQUATES SECTION

```
TMR0        EQU         1           ;means TMR0 is file 1.
STATUS      EQU         3           ;means STATUS is file 3.
PORTA       EQU         5           ;means PORTA is file 5.
PORTB       EQU         6           ;means PORTB is file 6.
ZEROBIT     EQU         2           ;means ZEROBIT is bit 2.
COUNT       EQU         0CH         ;means COUNT is file 0C,
                                    ;a register to count events.
;****************************************************

LIST        P=16F84     ;we are using the 16F84.
ORG         0           ;the start address in memory is 0
GOTO        START       ;goto start!

;****************************************************

;SUBROUTINE SECTION.

;3/32 second delay.
DELAY       CLRF        TMR0              ;START TMR0.
LOOPA       MOVF        TMR0,W            ;READ TMR0 INTO W.
            SUBLW       .3                ;TIME - 3
            BTFSS       STATUS,ZEROBIT    ;Check TIME-W = 0
            GOTO        LOOPA             ;Time is not = 3.
            RETLW       0                 ;Time is 3, return.

;****************************************************

;CONFIGURATION SECTION

START       BSF         STATUS,5          ;Turns to Bank1.
            MOVLW       B'00011111'       ;5 bits of PORTA are I/P
            TRIS        PORTA
            MOVLW       B'00000000'
            TRIS        PORTB             ;PORTB is OUTPUT
```

```
            MOVLW    B'00000111'    ;Prescaler is /256
            OPTION                  ;TIMER is 1/32 secs.
            BCF      STATUS,5       ;Return to Bank0.
            CLRF     PORTA          ;Clears PortA.
            CLRF     PORTB          ;Clears PortB.
```

;***
;
;Program starts now.
```
BEGIN       MOVLW    .10
            MOVWF    COUNT          ;Put 10 into COUNT.
PRESS       BTFSC    PORTA,0        ;Check switch is pressed
            GOTO     PRESS
            CALL     DELAY          ;Wait for 3/32 seconds.
            BCF      PORTB,0        ;TURN OFF LED.
RELEASE     BTFSS    PORTA,0        ;Check switch is released.
            GOTO     RELEASE
            CALL     DELAY          ;WAIT for 3/32 seconds.
            DECFSZ   COUNT          ;Dec COUNT skip if 0.
            GOTO     PRESS          ;Wait for another press.
            BSF      PORTB,0        ;Turn on LED.
            GOTO     BEGIN          ;Restart
```

END

How does it work?

- The file COUNT is first loaded with the count, i.e. 10, with:

```
            MOVLW    .10
            MOVWF    COUNT          ;Put 10 into COUNT.
```

- We then wait for the switch to be pressed, by PORTA,0 going low:

```
PRESS       BTFSC    PORTA,0        ;Check switch is pressed
            GOTO     PRESS
```

- Antibounce:

```
CALL        DELAY    ;Wait for 3/32 seconds.
```

- Turn off the LED on B0:

```
BCF         PORTB,0
```

● Wait for switch to be released

RELEASE	BTFSS	PORTA,0	;Check switch is released.
	GOTO	RELEASE	

● Antibounce:

CALL	DELAY	;Wait for 3/32 seconds.

● Decrement the file COUNT; if zero turn on LED and return to begin. If not zero continue pressing the switch.

DECFSZ	COUNT	;Dec COUNT skip if 0.
GOTO	PRESS	;Wait for another press.
BSF	PORTB,0	;Turn on LED.
GOTO	BEGIN	;Restart

This may appear to be a lot of programming to count presses of a switch, but once saved as a subroutine it can be reused in other programs.

Look up table

A look up table is used to change data from one form to another, i.e. pounds to kilograms, °C to °F, inches to centimetres, etc. The explanation of the operation of a look up table is best understood by way of an example.

7-segment display

Design a circuit that will count and display on a 7-segment display, the number of times a button is pressed, up to 10. The circuit diagram for this is shown in Figure 5.3.

The flowchart for the 7-segment display driver is shown in Figure 5.4

This is a basic solution that has a few omissions:

● The switch bounces when pressed.
● Clear the count at the start.
● The micro counts in binary, we require a 7-segment decimal display. So we need to convert the binary count to drive the relevant segments on the display.
● When the switch is released it bounces.

The amended flowchart is shown in Figure 5.5.

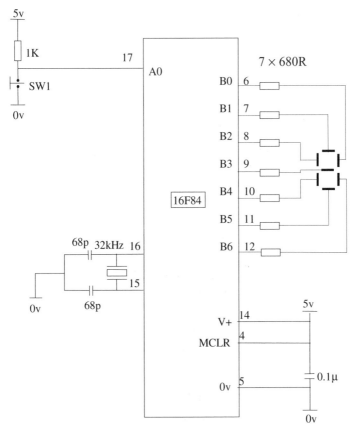

Figure 5.3 Circuit diagram of the 7-segment display driver

The flowchart is missing just one thing! What happens when the count reaches 10? The counter needs resetting (it would count up to 255 before resetting). The final flowchart is shown in Figure 5.7.

Now about this look up table. Figure 5.6 shows the configuration of PORTB to drive the 7-segment display (refer also to Figure 5.3).

The look up table for this is:

```
CONVERT     ADDWF    PC
            RETLW    B'01110111'    ;0
            RETLW    B'01000001'    ;1
            RETLW    B'00111011'    ;2
            RETLW    B'01101011'    ;3
```

```
RETLW       B'01001101'          ;4
RETLW       B'01101110'          ;5
RETLW       B'01111100'          ;6
RETLW       B'01000011'          ;7
RETLW       B'01111111'          ;8
RETLW       B'01001111'          ;9
```

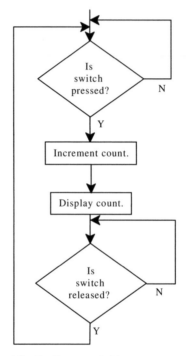

Figure 5.4 Initial Flowchart for the 7-segment driver

How does the look up table work?

Suppose we need to display a 0. We move 0 into W and CALL the look up table, here it is called CONVERT. The first line says ADD W to the Program Count, since W = 0 then goto the next line of the program which will return with the 7-segment value 0.

Suppose we need to display a 6. Move 6 into W and CALL CONVERT. The first line says ADD W to the Program Count, since W contains 6 then goto the next line of the program and move down six more lines and return with the code for 6, etc. Just one more thing: to check that a count has reached 10, subtract 10 from the count if the answer is 0, bingo!

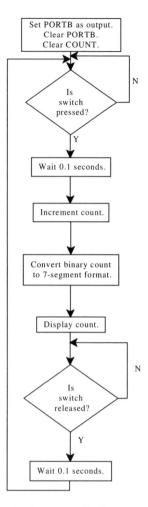

Figure 5.5 Amended flowchart for 7-segment display

NUMBER	PORTB							
	B7	B6	B5	B4	B3	B2	B1	B0
0	0	1	1	1	0	1	1	1
1	0	1	0	0	0	0	0	1
2	0	0	1	1	1	0	1	1
3	0	1	1	0	1	0	1	1
4	0	1	0	0	1	1	0	1
5	0	1	1	0	1	1	1	0
6	0	1	1	1	1	1	0	0
7	0	1	0	0	0	0	1	1
8	0	1	1	1	1	1	1	1
9	0	1	0	0	1	1	1	1

Figure 5.6 Binary code to drive 7-segment display

The program listing for the complete program

;DISPLAY.ASM

;EQUATES SECTION

```
TMR0        EQU     1          ;means TMR0 is file 1.
PC          EQU     2          ;Program Counter is file 2.
STATUS      EQU     3          ;means STATUS is file 3.
PORTA       EQU     5          ;means PORTA is file 5.
PORTB       EQU     6          ;means PORTB is file 6.
ZEROBIT     EQU     2          ;means ZEROBIT is bit 2.
COUNT       EQU     0CH        ;means COUNT is file 0C,
                              ;a register to count events.
;********************************************************

LIST        P=16F84            ;we are using the 16F84.
ORG         0                  ;the start address in memory is 0
GOTO        START              ;goto start!

;********************************************************

;SUBROUTINE SECTION.

;3/32 second delay.
DELAY       CLRF    TMR0            ;START TMR0.
LOOPA       MOVF    TMR0,W          ;READ TMR0 INTO W.
            SUBLW   .3              ;TIME - 3
            BTFSS   STATUS,ZEROBIT  ;Check TIME-W = 0
            GOTO    LOOPA           ;Time is not = 3.
            RETLW   0               ;Time is 3, return.

CONVERT     ADDWF   PC
            RETLW   B'01110111'     ;0
            RETLW   B'01000001'     ;1
            RETLW   B'00111011'     ;2
            RETLW   B'01101011'     ;3
            RETLW   B'01001101'     ;4
            RETLW   B'01101110'     ;5
            RETLW   B'01111100'     ;6
            RETLW   B'01000011'     ;7
            RETLW   B'01111111'     ;8
            RETLW   B'01001111'     ;9

;********************************************************
```

;CONFIGURATION SECTION

```
START    BSF      STATUS,5              ;Turns to Bank1.
         MOVLW    B'00011111'           ;5 bits of PORTA are I/P
         TRIS     PORTA
         MOVLW    B'00000000'
         TRIS     PORTB                 ;PORTB is OUTPUT
         MOVLW    B'00000111'           ;Prescaler is /256
         OPTION                         ;TIMER is 1/32 secs.
         BCF      STATUS,5              ;Return to Bank0.
         CLRF     PORTA                 ;Clears PortA.
         CLRF     PORTB                 ;Clears PortB.
;******************************************************
;Program starts now.


         CLRF     COUNT                 ;Set COUNT to 0.
PRESS    BTFSC    PORTA,0               ;Test for switch press.
         GOTO     PRESS                 ;Not pressed.
         CALL     DELAY                 ;Antibounce wait 0.1sec.
         INCF     COUNT                 ;Add 1 to COUNT.
         MOVF     COUNT,W               ;Move COUNT to W.
         SUBLW    .10                   ;COUNT-10, W is altered.
         BTFSC    STATUS,ZEROBIT        ;Is COUNT - 10 = 0?
         CLRF     COUNT                 ;Count = 10 Make Count=0
         MOVF     COUNT,W               ;Put Count in W again.
         CALL     CONVERT               ;Count is not 10, carry on.
         MOVWF    PORTB                 ;Output number to display.

RELEASE  BTFSS    PORTA,0               ;Is switch released?
         GOTO     RELEASE               ;Not released.
         CALL     DELAY                 ;Antibounce wait 0.1sec.
         GOTO     PRESS                 ;Look for another press.

END
```

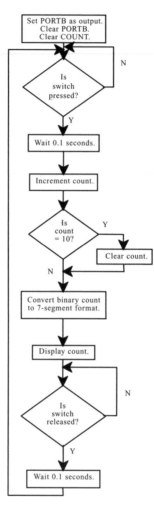

Figure 5.7 Final flowchart for 7-segment display

How does the program work?

- The file COUNT is cleared (to zero) and we wait for the switch to be pressed.

```
            CLRF        COUNT           ;Set COUNT to 0.
PRESS       BTFSC       PORTA,0         ;Test for switch press.
            GOTO        PRESS           ;Not pressed.
```

- Wait for 0.1 seconds, antibounce.

```
CALL        DELAY
```

- Add 1 to COUNT and check to see if it is 10:

INCF	COUNT	;Add 1 to COUNT.
MOVF	COUNT,W	;Move COUNT to W.
SUBLW	.10	;COUNT-10, W is altered.
BTFSC	STATUS,ZEROBIT	;Is COUNT - 10 = 0?

- If COUNT is 10, Clear it to 0 and output the count as 0. If the COUNT is not 10 then output the count.

CLRF	COUNT	;Count = 10 Make Count = 0
MOVF	COUNT,W	;Put Count in W again.
CALL	CONVERT	;Count is not 10, carry on.
MOVWF	PORTB	;Output number to display.

- Wait for the switch to be released and de-bounce. Then return to monitor the presses.

RELEASE	BTFSS	PORTA,0	;Is switch released?
	GOTO	RELEASE	;Not released.
	CALL	DELAY	;Antibounce wait 0.1 sec.
	GOTO	PRESS	;Look for another press.

Test your understanding:

- Modify the program to count up to 6 and reset.
- Modify the program to count up to F in hex and reset.

A look up table to change °C to °F called DEGREE, is shown below.

DEGREE	ADDWF	PC	;ADD W to Program Count.
	RETLW	.32	;0°C = 32°F
	RETLW	.34	;1°C = 34°F
	RETLW	.36	;2°C = 36°F
	RETLW	.37	;3°C = 37°F
	RETLW	.39	;4°C = 39°F
	RETLW	.41	;5°C = 41°F
	RETLW	.43	;6°C = 43°F
	RETLW	.45	;7°C = 45°F
	RETLW	.46	;8°C = 46°F
	RETLW	.48	;9°C = 48°F
	RETLW	.50	;10°C = 50°F
	RETLW	.52	;11°C = 52°F

RETLW	.54	;12°C = 54°F
RETLW	.55	;13°C = 55°F
RETLW	.57	;14°C = 57°F
RETLW	.59	;15°C = 59°F
RETLW	.61	;16°C = 61°F
RETLW	.63	;17°C = 63°F
RETLW	.64	;18°C = 64°F
RETLW	.66	;19°C = 66°F
RETLW	.68	;20°C = 68°F
RETLW	.70	;21°C = 70°F
RETLW	.72	;22°C = 72°F
RETLW	.73	;23°C = 73°F
RETLW	.75	;24°C = 75°F
RETLW	.77	;25°C = 77°F
RETLW	.79	;26°C = 79°F
RETLW	.81	;27°C = 81°F
RETLW	.82	;28°C = 82°F
RETLW	.84	;29°C = 84°F
RETLW	.86	;30°C = 86°F

Another application of the use of the look up table is a solution for a previous example, i.e. the 'Control application – a hot air blower', introduced in Chapter 3.

In this example when PORTA was read the data was treated as a binary number, but we could just as easily treat the data as a decimal number, i.e.

A2 A1 A0   = 000 or 0
             = 001 or 1
             = 010 or 2
             = 011 or 3
             = 100 or 4
             = 101 or 5
             = 110 or 6
             = 111 or 7

The look up table for this would be:

CONVERT	ADDWF	PC	
	RETLW	B'00000010'	;0 on PORTA turns on B1
	RETLW	B'00000001'	;1 on PORTA turns on B0
	RETLW	B'00000011'	;2 on PORTA turns on B1,B0
	RETLW	B'00000001'	;3 on PORTA turns on B0
	RETLW	B'00000000'	;4 on PORTA turns off B1,B0

```
RETLW       B'00000001'        ;5 on PORTA turns on B0
RETLW       B'00000000'        ;6 on PORTA turns off B1,B0
RETLW       B'00000010'        ;7 on PORTA turns on B1
```

The complete program listing for the program DISPLAY2

;DISPLAY2.ASM

;EQUATES SECTION

```
PC        EQU     2         ;Program Counter is file 2.
STATUS    EQU     3         ;means STATUS is file 3.
PORTA     EQU     5         ;means PORTA is file 5.
PORTB     EQU     6         ;means PORTB is file 6.
COUNT     EQU     0CH       ;means COUNT is file 0C,
                            ;a register to count events.
;************************************************************
LIST      P=16F84           ;we are using the 16F84.
ORG       0                 ;the start address in memory is 0
GOTO      START             ;goto start!

;************************************************************

;SUBROUTINE SECTION.

CONVERT   ADDWF    PC
          RETLW    B'00000010'    ;0 on PORTA turns on B1
          RETLW    B'00000001'    ;1 on PORTA turns on B0
          RETLW    B'00000011'    ;2 on PORTA turns on B1,B0
          RETLW    B'00000001'    ;3 on PORTA turns on B0
          RETLW    B'00000000'    ;4 on PORTA turns off B1,B0
          RETLW    B'00000001'    ;5 on PORTA turns on B0
          RETLW    B'00000000'    ;6 on PORTA turns off B1,B0
          RETLW    B'00000010'    ;7 on PORTA turns on B1

;************************************************************

;CONFIGURATION SECTION

START     BSF      STATUS,5       ;Turns to Bank1.
          MOVLW    B'00011111'    ;5 bits of PORTA are I/P
          TRIS     PORTA
          MOVLW    B'00000000'
          TRIS     PORTB          ;PORTB is OUTPUT
```

```
                 BCF        STATUS,5        ;Return to Bank0.
                 CLRF       PORTA           ;Clears PortA.
                 CLRF       PORTB           ;Clears PortB.
```

;**
;

;Program starts now.

```
BEGIN            MOVF       PORTA,W         ;Read PORTA into W
                 CALL       CONVERT         ;Obtain O/Ps from I/Ps.
                 MOVWF      PORTB           ;switch on O/Ps
                 GOTO       BEGIN           ;repeat
END
```

How does the program work?

- The program first of all reads the value of PORTA into the working register, W:

```
                 MOVF       PORTA,W
```

- The CONVERT routine is called which returns with the correct setting of the outputs in W, i.e. if the value of PORTA was 3 then the look up table would return with 00000001 in W to turn on B0 and turn off B1:

```
                 CALL       CONVERT         ;Obtain O/Ps from I/Ps.
                 MOVWF      PORTB           ;switch on O/Ps
```

- The program then returns to check the setting of PORTA again.

Numbers larger than 255

The PIC Microcontrollers are 8-bit devices, this means that they can easily count up to 255 using one memory location. But to count higher then more than one memory location has to be used for the count. Consider counting a switch press up to 1000 and then turn on an LED to show this count has been achieved. The circuit for this is shown in Figure 5.8.

To count up to 1000 in decimal, i.e. 03E8 in hex, files COUNTB and COUNTA will store the count (a count of 65 535 is then possible). COUNTB will count up to 03H, then when COUNTA has reached E8H, LED1 will light indicating the count of 1000 has been reached. The flowchart for this 1000 count is shown in Figure 5.9.

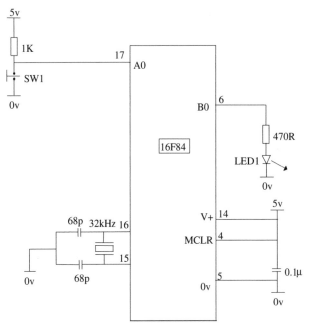

Figure 5.8 Circuit for 1000 count

Flowchart explanation

- The program is waiting for SW1 to be pressed. When it is, there is a delay of 0.1 seconds to allow the switch bounce to stop.
- The program then looks for the switch to be released and waits 0.1 seconds for the bounce to stop.
- 1 is then added to COUNTA and a check is made to see if the count has overflowed, i.e. reached 256. (255 is the maximum it will hold; when it reaches 256 it will reset to zero as a two digit counter would reset to zero going from 99 to 100.)
- If COUNTA has overflowed then we increment COUNTB.
- A check is made to see if COUNTB has reached 03H, if not we return to keep counting.
- If COUNTB has reached 03H then we count presses until COUNTA reaches E8H. The count in decimal is then 1000 and the LED is lit.

Any count can be attained by altering the values COUNTB and COUNTA are allowed to count up to, i.e. to count up to 5000 in decimal which is 1388H. Ask if COUNTB = 13H then count until COUNTA has reached 88H.

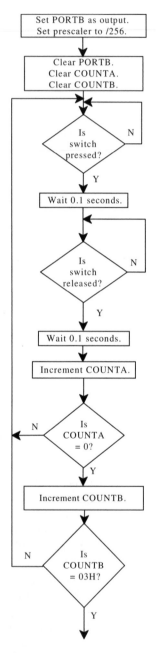

Figure 5.9 1000 count flowchart

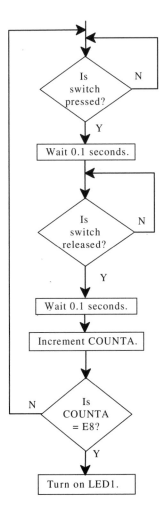

The program listing
;CNT1000.ASM

;EQUATES SECTION

TMR0	EQU	1	;means TMR0 is file 1.
STATUS	EQU	3	;means STATUS is file 3.
PORTA	EQU	5	;means PORTA is file 5.
PORTB	EQU	6	;means PORTB is file 6.
ZEROBIT	EQU	2	;means ZEROBIT is bit 2.
COUNTA	EQU	0CH	;USER RAM LOCATION.
COUNTB	EQU	0DH	;USER RAM LOCATION.

;***

LIST	P=16F84	;we are using the 16F84.
ORG	0	;the start address in memory is 0
GOTO	START	;goto start!

;***

;SUBROUTINE SECTION.

;3/32 second delay.

DELAY	CLRF	TMR0	;START TMR0.
LOOPA	MOVF	TMR0,W	;READ TMR0 INTO W.
	SUBLW	.3	;TIME - 3
	BTFSS	STATUS,ZEROBIT	;Check TIME-W = 0
	GOTO	LOOPA	;Time is not = 3.
	RETLW	0	;Time is 3, return.

;***

;CONFIGURATION SECTION

START	BSF	STATUS,5	;Turns to Bank1.
	MOVLW	B'00011111'	;5 bits of PORTA are I/P
	TRIS	PORTA	
	MOVLW	B'00000000'	
	TRIS	PORTB	;PORTB is OUTPUT
	MOVLW	B'00000111'	;Prescaler is /256
	OPTION		;TIMER is 1/32 secs.
	BCF	STATUS,5	;Return to Bank0.

```
          CLRF        PORTA           ;Clears PortA.
          CLRF        PORTB           ;Clears PortB.
;*******************************************************
;Program starts now.


          CLRF        COUNTA
          CLRF        COUNTB

PRESS     BTFSC       PORTA,0         ;Check switch pressed
          GOTO        PRESS
          CALL        DELAY           ;Wait for 3/32 seconds.
RELEASE   BTFSS       PORTA,0         ;Check switch is released.
          GOTO        RELEASE
          CALL        DELAY           ;Wait for 3/32 seconds.
          INCFSZ      COUNTA          ;Inc. COUNT skip if 0.
          GOTO        PRESS

          INCF        COUNTB
          MOVLW       03H             ;Put 03H in W.   *
          SUBWF       COUNTB,W        ;COUNTB - W (i.e. 03)
          BTFSS       STATUS,ZEROBIT  ;IS COUNTB=03H
          GOTO        PRESS           ;No

PRESS1    BTFSC       PORTA,0         ;Check switch pressed.
          GOTO        PRESS1
          CALL        DELAY           ;Wait for 3/32 seconds.
RELEASE1  BTFSS       PORTA,0         ;Check switch released.
          GOTO        RELEASE1
          CALL        DELAY           ;Wait for 3/32 seconds.
          INCF        COUNTA
          MOVLW       0E8H            ;Put E8 in W.   *
          SUBWF       COUNTA          ;COUNTA – E8.
          BTFSS       STATUS,ZEROBIT  ;COUNTA=E8?
          GOTO        PRESS1          ;No.
          BSF         PORTB,0         ;Yes, turn on LED1.
STOP      GOTO        STOP            ;stop here
END
```

How does the program work?

- The two files used for counting are cleared:

```
          CLRF      COUNTA
          CLRF      COUNTB
```

● As we have done previously we wait for the switch to be pressed and released and to stop bouncing:

```
PRESS     BTFSC     PORTA,0          ;Check switch pressed
          GOTO      PRESS
          CALL      DELAY            ;Wait for 3/32 seconds.
RELEASE   BTFSS     PORTA,0          ;Check switch is released.
          GOTO      RELEASE
          CALL      DELAY            ;Wait for 3/32 seconds.
```

● We add 1 to file COUNTA and check to see if it is zero. If it isn't then continue monitoring presses. (The file would be zero when we add 1 to the 8-bit number 1111 1111, and it overflows to 0000 0000):

```
          INCFSZ    COUNTA           ;Inc. COUNT skip if 0.
          GOTO      PRESS
```

● If the file COUNTA has overflowed then we add 1 to the file COUNTB, just as you would do with two columns of numbers. We then need to know if COUNTB has reached 03H. If COUNTB is not 03H then we return to PRESS and continue monitoring the presses:

```
          INCF      COUNTB
          MOVLW     03H              ;Put 03H in W.
          SUBWF     COUNTB,W         ;COUNTB - W (i.e. 03)
          BTFSS     STATUS,ZEROBIT   ;IS COUNTB=03H?
          GOTO      PRESS            ;No
```

● Once COUNTB has reached 03H we need only wait until COUNTA reaches 0E8H and we would have counted up to 03E8H, i.e. 5000 in decimal. Then we turn on the LED:

```
PRESS1    BTFSC     PORTA,0          ;Check switch pressed.
          GOTO      PRESS1
          CALL      DELAY            ;Wait for 3/32 seconds.
RELEASE1  BTFSS     PORTA,0          ;Check switch released.
          GOTO      RELEASE1
          CALL      DELAY            ;Wait for 3/32 seconds.
          INCF      COUNTA
          MOVLW     0E8H             ;Put E8 in W.
          SUBWF     COUNTA           ;COUNTA – E8.
```

```
            BTFSS    STATUS,ZEROBIT    ;COUNTA=E8?
            GOTO     PRESS1            ;No.
            BSF      PORTB,0           ;Yes, turn on LED1.
STOP        GOTO     STOP              ;stop here
```

This listing can be used as a subroutine in your program to count up to any number to 65 535 (or more if you use a COUNTC file). Just alter COUNTB and COUNTA to whatever values you wish, in the two places marked * in the program.

Question: How would you count up to 20 000?

Answer: (Have you tried it first!) 20 000 = 4E20H so COUNTB would count up to 4EH and COUNTA would then count to 20H.

Question: How would you count to 100 000?

Answer: 100 000 = 0186A0H, you would use a third file COUNTC to count to 01H, COUNTB would count to 86H and COUNTA would count to A0H.

Programming can be made a lot simpler by keeping a library of subroutines. Here is another . . .

Long time intervals

Probably the more frequent use of a large count is to count TMR0 pulses to generate long time intervals. We have previously seen in the DELAY Subroutine that we can slow the internal timer clock down to 1/32 seconds. Counting a maximum of 255 of these gives a time of 255 × 1/32 = 8 seconds. Suppose we want to turn on an LED for 5 minutes when a switch is pressed. 5 minutes = 300 seconds = 300 × 32 (1/32 seconds), i.e. a TMR0 count of 9600. This is 2580 in hex. The circuit is the same as for the 1000 count circuit, and the flowchart is shown in Figure 5.10.

Explanation of the flowchart

1 Wait until the switch is pressed, the LED is then turned on.
2 TMR0 is cleared to start the timing interval.
3 TMR0 is moved into W (read) to catch the first count.
4 Then wait for TMR0 to return to zero (the count will be 256), i.e. 100 in hex.
5 COUNTA is then incremented and steps 3 and 4 repeated until COUNTA reaches 25H.
6 Wait until TMR0 has reached 80H.
7 The count has reached 2580H, i.e. 9600 in decimal. 5 minutes has elapsed and the LED is turned off.

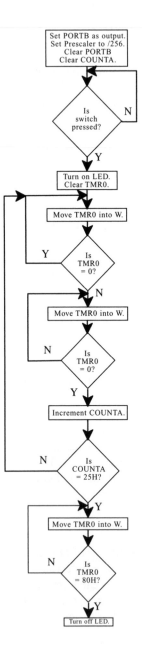

Figure 5.10 Flowchart for the 5 minute delay

Program listing for 5 minute delay

;LONGDLY.ASM

;EQUATES SECTION

TMR0	EQU	1	;means TMR0 is file 1.
STATUS	EQU	3	;means STATUS is file 3.
PORTA	EQU	5	;means PORTA is file 5.
PORTB	EQU	6	;means PORTB is file 6.
ZEROBIT	EQU	2	;means ZEROBIT is bit 2.
COUNTA	EQU	0CH	;USER RAM LOCATION.

;**

LIST	P=16F84		;we are using the 16F84.
ORG	0		;the start address in memory is 0
GOTO	START		;goto start!

;**

;CONFIGURATION SECTION

START	BSF	STATUS,5	;Turns to Bank1.
	MOVLW	B'00011111'	;5 bits of PORTA are I/P
	TRIS	PORTA	
	MOVLW	B'00000000'	
	TRIS	PORTB	;PORTB is OUTPUT
	MOVLW	B'00000111'	;Prescaler is /256
	OPTION		;TIMER is 1/32 secs.
	BCF	STATUS,5	;Return to Bank0.
	CLRF	PORTA	;Clears PortA.
	CLRF	PORTB	;Clears PortB.

;**
;Program starts now.

	CLRF	COUNTA	
PRESS	BTFSC	PORTA,0	;Check switch pressed.
	GOTO	PRESS	;No
	BSF	PORTB,0	;Yes, turn on LED
	CLRF	TMR0	;Start TMR0.
WAIT0	MOVF	TMR0,W	;Move TMR0 into W
	BTFSC	STATUS,ZEROBIT	;Is TMR0 = 0.
	GOTO	WAIT0	;Yes

```
WAIT1       MOVF        TMR0,W          ;No, move TMR0 into W.
            BTFSS       STATUS,ZEROBIT
            GOTO        WAIT1           ;Wait for TMR0 to overflow
            INCF        COUNTA          ;Increment COUNTA
            MOVLW       25H
            SUBWF       COUNTA,W        ;COUNTA - 25H
            BTFSS       STATUS,ZEROBIT  ;Is COUNTA = 25H
            GOTO        WAIT0           ;COUNTA < 25H
WAIT2       MOVF        TMR0,W          ;COUNTA = 25H
            MOVLW       80H
            SUBWF       TMR0,W          ;TMR0 - 80H
            BTFSS       STATUS,ZEROBIT  ;Is TMR0=80H
            GOTO        WAIT2           ;TMR0 < 80H
            BCF         PORTB,0         ;TMR0=80H, turn off LED
END
```

The explanation of this program operation is similar to that of the count to 1000, done earlier in this chapter. This listing can be used as a subroutine and times up to 65 535 × 1/32 seconds, i.e. 34 minutes, can be obtained.

Problem: Change the listing to produce a 30 minute delay.

Hint: 1800 seconds in hex is 0708H.

One hour delay

Another and probably simpler way of obtaining a delay of, say 1 hour, is:

• Write a delay of 5 seconds.
• CALL it 6 times, this gives a delay of 30 seconds.
Put this in a loop to repeat 120 times, i.e. 120 × 30 seconds = 1 hour.

The code for the 1 hour subroutine will look like:

```
ONEHOUR     MOVLW       .120        ;put 120 in W
            MOVWF       COUNT       ;load COUNT with 120
LOOP        CALL        DELAY5      ;Wait 5 seconds
            CALL        DELAY5      ;Wait 5 seconds
            CALL        DELAY5      ;Wait 5 seconds
            CALL        DELAY5      ;Wait 5 seconds
            CALL        DELAY5      ;Wait 5 seconds
            CALL        DELAY5      ;Wait 5 seconds
            DECFSZ      COUNT       ;Subtract 1 from COUNT
```

```
        GOTO      LOOP      ;Count is not zero.
        RETLW     0         ;RETURN to program.
```

The Program for the 1 hour delay

```
; ONEHOUR.ASM for 16F84.   This sets PORTA as an INPUT (NB 1
;                          means input) and PORTB as an OUTPUT
;                          (NB 0 means output). The OPTION
;                          Register is set to /256 to give timing pulses
;                          of 1/32 of a second.
;                          1 hour and 5 second delays are
;                          included in the subroutine section.

;********************************************************

;EQUATES SECTION

TMR0      EQU      1        ;means TMR0 is file 1.
STATUS    EQU      3        ;means STATUS is file 3.
PORTA     EQU      5        ;means PORTA is file 5.
PORTB     EQU      6        ;means PORTB is file 6.
ZEROBIT   EQU      2        ;means ZEROBIT is bit 2.
COUNT     EQU      0CH      ;means COUNT is file 0C,
                           ;a register to count events.
;********************************************************
LIST      P=16F84           ;we are using the 16F84.
ORG       0                 ;the start address in memory is 0
GOTO      START             ;goto start!

;********************************************************

;SUBROUTINE SECTION.

;1 hour delay.
ONEHOUR   MOVLW    .120     ;put 120 in W
          MOVWF    COUNT    ;load COUNT with 120
LOOP      CALL     DELAY5   ;Wait 5 seconds
          CALL     DELAY5   ;Wait 5 seconds
          CALL     DELAY5   ;Wait 5 seconds
          CALL     DELAY5   ;Wait 5 seconds
          CALL     DELAY5   ;Wait 5 seconds
          CALL     DELAY5   ;Wait 5 seconds
          DECFSZ   COUNT    ;Subtract 1 from COUNT
          GOTO     LOOP     ;Count is not zero.
```

```
                    RETLW      0                      ;RETURN to program.

;5 second delay.
DELAY5              CLRF       TMR0                   ;START TMR0.
LOOPB               MOVF       TMR0,W                 ;READ TMR0 INTO W.
                    SUBLW      .160                   ;TIME - 160
                    BTFSS      STATUS,ZEROBIT         ;Check TIME-W = 0
                    GOTO       LOOPB                  ;Time is not = 160.
                    RETLW      0                      ;Time is 160, return.
```

;**

;CONFIGURATION SECTION

```
START               BSF        STATUS,5               ;Turns to Bank1.
                    MOVLW      B'00011111'            ;5 bits of PORTA are I/P
                    TRIS       PORTA
                    MOVLW      B'00000000'
                    TRIS       PORTB                  ;PORTB is OUTPUT
                    MOVLW      B'00000111'            ;Prescaler is /256
                    OPTION                            ;TIMER is 1/32 secs.
                    BCF        STATUS,5               ;Return to Bank0.
                    CLRF       PORTA                  ;Clears PortA.
                    CLRF·      PORTB                  ;Clears PortB.
```

;**

;Program starts now.
```
                    BSF        PORTB,0                ;Turn on B0
                    CALL       ONEHOUR                ;Wait 1 Hour.
                    BCF        PORTB,0                ;Turn off B0.
STOP                GOTO       STOP                   ;STOP!
END
```

6
The 16C54 microcontroller

The 16C54 is an example of a One Time Programmable (OTP) device; it was brought out before the 16F84. The main difference between them is that the 16C54 is not electrically erasable, it has to be erased by UV light for about 15 minutes.

The 16C54JW version is UV erasable and the 16C54LP is a One Time (only) Programmable (OTP), 32kHz version. You would use a 16C54JW for development and then program an OTP device for your final circuit. The OTP device has to be selected for the correct oscillator, i.e. LP for 32kHz crystal, XT for 4MHz, HS for 20MHz and R-C for an R-C network.

The header for use with the 16C54 is shown below.

Header for the 16C54

```
;HEADER54.ASM for 16C54.  This sets PORTA as an INPUT (NB 1
;                            means input) and PORTB as an OUTPUT
;                            (NB 0 means output). The OPTION
;                            register is set to /256 to give timing pulses
;                            of 1/32 of a second.
;                            1 second and 0.5 second delays are
;                            included in the subroutine section.

;********************************************************

;EQUATES SECTION

TMR0        EQU        1        ;means TMR0 is file 1.
STATUS      EQU        3        ;means STATUS is file 3.
PORTA       EQU        5        ;means PORTA is file 5.
PORTB       EQU        6        ;means PORTB is file 6.
ZEROBIT     EQU        2        ;means ZEROBIT is bit 2.
COUNT       EQU        7        ;means COUNT is file 7,
                                ;a register to count events
TIME        EQU        8        ;file 8 where the time is stored.
;********************************************************
```

```
LIST        P=16C54       ;we are using the 16C54.
ORG         01FFH         ;the start address in memory is 1FF at the end.
GOTO        START         ;goto start!
ORG         0
```

;***

;SUBROUTINE SECTION.

;1 second delay.
```
DELAY1      CLRF    TMR0              ;START TMR0.
LOOPA       MOVLW   .32
            MOVWF   TIME              ;Time = 32/32 secs.
            MOVF    TMR0,W            ;Read TMR0 into W.
            SUBWF   TIME,W            ;TIME - 32, result in W.
            BTFSS   STATUS,ZEROBIT    ;Check TIME-W = 0
            GOTO    LOOPA             ;Time is not = 32.
            RETLW   0                 ;Time is 32, return.
```

;0.5 second delay.
```
DELAYP5     CLRF    TMR0              ;START TMR0.
LOOPB       MOVLW   .16
            MOVWF   TIME              ;Time = 16/32 secs.
            MOVF    TMR0,W            ;READ TMR0 INTO W.
            SUBWF   TIME,W            ;TIME - 16
            BTFSS   STATUS,ZEROBIT    ;Check TIME-W = 0
            GOTO    LOOPB             ;Time is not = 16.
            RETLW   0                 ;Time is 16, return.
```

;***

;CONFIGURATION SECTION

```
START       MOVLW   B'00001111'       ;4 bits of PORTA are I/P
            TRIS    PORTA
            MOVLW   B'00000000'
            TRIS    PORTB             ;PORTB is OUTPUT
            MOVLW   B'00000111'       ;Prescaler is /256
            OPTION                    ;TIMER is 1/32 secs.
            CLRF    PORTA             ;Clears PortA.
            CLRF    PORTB             ;Clears PortB.
```

;***
;Program starts now.

This header can now be used to write programs for the 16C54 microcontroller.

There are a number of differences between the 16F84 and the 16C54 that the header has taken care of, but be aware of the differences when writing your program.

- The 16C54 does not use Banks so there is no need to change from one to the other.
- There are only seven registers on the 16C54 (see 16C54 memory map, Figure 6.1). So the user files start at number 7, i.e. COUNT EQU 7, TIME EQU 8.
- The 16C54 does not have the instruction SUBLW. So in the DELAY subroutine the delay is moved into a file called TIME. (Note: TIME EQUATES TO 8.) Then the delay in the file is subtracted from W, giving the same result as for the 16F84.
- Why bother using the 16C54? The reprogrammable 16C54, i.e. the 16C54JW, is more expensive than the 16F84. But the One Time Programmable (OTP) 16C54, i.e. the 16C54/04P, is cheaper. So when your design is final you can blow the program into the cheaper 16C54/04P. Why bother with the expensive 16C54JW and not the 16F84 for program development? Only convenience – not having to change the program.
- The 16C54JW has to be erased under an ultraviolet lamp for about 15 minutes – this is a bind if you are impatient, you may need a couple.
- Pin 3 is only a T0CKI pin, it does not double as A4 like the 16F84 and must be pulled high if the T0CKI is not being used.

16C54 memory map

FILE ADDRESS	FILENAME
00	INDIRECT ADDRESS
01	TMR0
02	PC
03	STATUS
04	FSR
05	PORT A
06	PORTB
07	USER FILE
08	USER FILE
09	USER FILE
0A	USER FILE
0B	USER FILE
0C	USER FILE
0D	USER FILE
0E	USER FILE
0F	USER FILE
10	USER FILE
11	USER FILE
12	USER FILE
13	USER FILE
14	USER FILE
15	USER FILE
16	USER FILE
17	USER FILE
18	USER FILE
19	USER FILE
1A	USER FILE
1B	USER FILE
1C	USER FILE
1D	USER FILE
1E	USER FILE
1F	USER FILE

Figure 6.1 16C54 memory map

7
Alphanumeric displays

New instructions used in this chapter:
- NOP

Using an alphanumeric display in a project can bring it alive. Instead of showing a number on a 7-segment display the alphanumeric display could indicate 'The Temperature is 27°C'. Instructions can also be given on screen.

This section details the use of a 16 character by 2 line display, which incorporates an Hitachi HD44780 liquid crystal display controller driver chip. The HD44780 is an industry standard that is also used in displays other than Hitachi (fortunately). The chip is also used as a driver for other display configurations, i.e. 16×1, 20×2, 20×4, 40×2, etc. It has an onboard character generator ROM which can display 192 character patterns.

The circuit diagram connecting the alphanumeric display to the 16F84 is shown in Figure 7.1. This configuration is for the HD44780 driver and can be used with any of the displays using this chip.

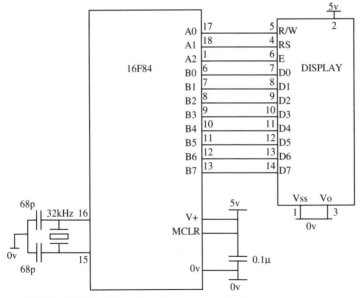

Figure 7.1 The 16F84 driving the alphanumeric display

Display pin identification

This display configuration shows 11 outputs from the microcontroller, three control lines and eight data lines connecting to the display. R/W is the read/write control line, RS is the register select and E is the chip enable. The R/W line tells the display to expect data to be written to it or to have data read from it. The data that is written to it is the address of the character, the code for the character or the type of command we require it to perform such as turn the cursor off.

The R/S line selects either a command to perform (R/S = 0), i.e. clear display, turn cursor on or off, or selects a data transfer (R/S = 1). The E line enables, (E = 1) and disables, (E = 0) the display.

There is much more to this display than we are able to look at here. If you wish to know more about them you will need to consult the manufacturer's data book.

If we use 11 lines to drive the display that would only leave two lines for the rest of our control with the 16F84. We could of course use a micro with 22 or 33 I/O. The display can, however, be driven with four data lines instead of eight, 4 bits of data are then sent twice. This complicates the program a little – but since I have done that work in the header it requires no more effort on your part.

Also the R/W line is used to write commands to the micro and read the busy line which indicates when the relatively slow display has processed the data. If we allow the micro enough time to complete its task then we do not have to read the busy line, we can just write to the display. The R/W line can then be connected to 0v in a permanent write mode and we do not require a read/write line from the micro.

We will therefore only require four data lines and two control lines to drive the display leaving seven lines available for I/O on the 16F84.

This six line control for the display is shown in Figure 7.2.

Configuring the display

Before writing to the display you first have to configure it. That means tell it if you are:

- using a 4-bit or 8-bit microcontroller,
- using a one or two line display,
- using a character font size of 5 × 10 or 5 × 7 dots,
- turning the display on or off,

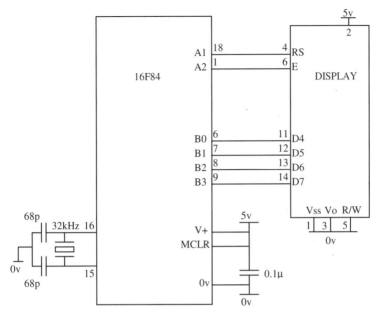

Figure 7.2 The 16F84 driving the alphanumeric display with six control lines

- turning the cursor on or off,
- incrementing the cursor or not. The cursor position increments after a character has been written to the display.

In the program shown below the display has been set up in the Configuration Section with Function Set at 32H to use a 4-bit microcontroller with a two line display and Font size of 5 × 7 dots. The Display is turned on and Cursor turned off with 0CH and the Cursor set to increment with 06H. This information was obtained from the display data sheet.

Writing to the display

- To write to the display you first set the address of the cursor (where you want the character to appear). The Cursor address locations are shown in Figure 7.3. Line1 address starts at 80H. Line2 address starts at C0H.

- Then tell the display what the character code is, e.g. A has the code 41H, B has the code 42H, C is 43H, 0 is 30H, 1 is 31H, 2 is 32H, etc.

To print an A on the screen – first enable the display, send 2 to PORTA, send the code 41H to PORTB and CLOCK this data.

These instructions have been written in the Subroutine Section so all you have to do is CALL A.

80	81	82	83	84	85	86	87	88	89	8A	8B	8C	8D	8E	8F
C0	C1	C2	C3	C4	C5	C6	C7	C8	C9	CA	CB	CC	CD	CE	CF

Figure 7.3 Cursor address location

To write HELLO on the display the program would be:

CALL H
CALL E
CALL L
CALL L
CALL O

Program example

The program below is the listing to spell out MICROCONTROLLERS AT THE MMU. Then CONTACT DAVE SMITH. Together with the time delays.

```
;ANHEAD84.ASM Header for the alphanumeric display using 6 I/O

         TMR0     EQU 1              ;TMR0 is FILE 1.
         PORTA    EQU 5              ;PORTA is FILE 5.
         PORTB    EQU 6              ;PORTB is FILE 6.
         STATUS   EQU 3              ;STATUS is FILE3.
         ZEROBIT  EQU 2              ;ZEROBIT is Bit 2.
;**********************************************************
         LIST     P=16F84            ;We are using the 16F84.
         ORG      0                  ;0 is the start address.
         GOTO     START              ;goto start!


;**********************************************************
;SUBROUTINE SECTION.

;3 SECOND DELAY
DELAY3   CLRF     TMR0               ;Start TMR0
LOOPA    MOVF     TMR0,W             ;Read TMR0 into W
         SUBLW    .96                ;TIME - W
         BTFSS    STATUS,ZEROBIT     ;Check TIME-W=0
         GOTO     LOOPA
         RETLW    0                  ;return after TMR0 = 96

;P1 SECOND DELAY
DELAYP1  CLRF     TMR0               ;Start TMR0
LOOPC    MOVF     TMR0,W             ;Read TMR0 into W
         SUBLW    3                  ;TIME - W
```

```
             BTFSS      STATUS,ZEROBIT   ;Check TIME-W=0
             GOTO       LOOPC
             RETLW      0                ;return after TMR0 = 3

CLOCK        BSF        PORTA,2
             NOP
             BCF        PORTA,2
             NOP
             RETLW      0
;*******************************************************
A            MOVLW      2                ;enables the display
             MOVWF      PORTA
             MOVLW      4H
             MOVWF      PORTB
             CALL       CLOCK
             MOVLW      1H               ;41 is code for A
             MOVWF      PORTB
             CALL       CLOCK            ;clock character onto display.
             RETLW      0

BB           MOVLW      2                ;enables the display
             MOVWF      PORTA
             MOVLW      4H
             MOVWF      PORTB
             CALL       CLOCK
             MOVLW      2H               ;42 is code for B
             MOVWF      PORTB
             CALL       CLOCK            ;clock character onto display.
             RETLW      0

C            MOVLW      2                ;enables the display
             MOVWF      PORTA
             MOVLW      4H
             MOVWF      PORTB
             CALL       CLOCK
             MOVLW      3H
             MOVWF      PORTB
             CALL       CLOCK            ;clock character onto display.
             RETLW      0

D            MOVLW      2                ;enables the display
             MOVWF      PORTA
             MOVLW      4H
             MOVWF      PORTB
```

```
          CALL    CLOCK
          MOVLW   4H
          MOVWF   PORTB
          CALL    CLOCK          ;clock character onto display.
          RETLW   0

E         MOVLW   2              ;enables the display
          MOVWF   PORTA
          MOVLW   4H
          MOVWF   PORTB
          CALL    CLOCK
          MOVLW   5H
          MOVWF   PORTB
          CALL    CLOCK          ;clock character onto display.
          RETLW   0

F         MOVLW   2              ;enables the display
          MOVWF   PORTA
          MOVLW   4H
          MOVWF   PORTB
          CALL    CLOCK
          MOVLW   6H
          MOVWF   PORTB
          CALL    CLOCK          ;clock character onto display.
          RETLW   0

G         MOVLW   2              ;enables the display
          MOVWF   PORTA
          MOVLW   4H
          MOVWF   PORTB
          CALL    CLOCK
          MOVLW   7H
          MOVWF   PORTB
          CALL    CLOCK          ;clock character onto display.
          RETLW   0

H         MOVLW   2              ;enables the display
          MOVWF   PORTA
          MOVLW   4H
          MOVWF   PORTB
          CALL    CLOCK
          MOVLW   8H
          MOVWF   PORTB
          CALL    CLOCK          ;clock character onto display.
          RETLW   0
```

```
I    MOVLW    2              ;enables the display
     MOVWF    PORTA
     MOVLW    4H
     MOVWF    PORTB
     CALL     CLOCK
     MOVLW    9H
     MOVWF    PORTB
     CALL     CLOCK          ;clock character onto display.
     RETLW    0

J    MOVLW    2              ;enables the display
     MOVWF    PORTA
     MOVLW    4H
     MOVWF    PORTB
     CALL     CLOCK
     MOVLW    0AH
     MOVWF    PORTB
     CALL     CLOCK          ;clock character onto display.
     RETLW    0

K    MOVLW    2              ;enables the display
     MOVWF    PORTA
     MOVLW    4H
     MOVWF    PORTB
     CALL     CLOCK
     MOVLW    0BH
     MOVWF    PORTB
     CALL     CLOCK          ;clock character onto display.
     RETLW    0

L    MOVLW    2              ;enables the display
     MOVWF    PORTA
     MOVLW    4H
     MOVWF    PORTB
     CALL     CLOCK
     MOVLW    0CH
     MOVWF    PORTB
     CALL     CLOCK          ;clock character onto display.
     RETLW    0

M    MOVLW    2              ;enables the display
     MOVWF    PORTA
     MOVLW    4H
     MOVWF    PORTB
```

```
              CALL     CLOCK
              MOVLW    0DH
              MOVWF    PORTB
              CALL     CLOCK          ;clock character onto display.
              RETLW    0

N             MOVLW    2              ;enables the display
              MOVWF    PORTA
              MOVLW    4H
              MOVWF    PORTB
              CALL     CLOCK          ;clock character onto display.
              MOVLW    0EH
              MOVWF    PORTB
              CALL     CLOCK          ;clock character onto display.
              RETLW    0

O             MOVLW    2              ;enables the display
              MOVWF    PORTA
              MOVLW    4H
              MOVWF    PORTB
              CALL     CLOCK
              MOVLW    0FH
              MOVWF    PORTB
              CALL     CLOCK          ;clock character onto display.
              RETLW    0

P             MOVLW    2
              MOVWF    PORTA
              MOVLW    5H
              MOVWF    PORTB
              CALL     CLOCK
              MOVLW    0H
              MOVWF    PORTB
              CALL     CLOCK          ;clock character onto display.
              RETLW    0

Q             MOVLW    2
              MOVWF    PORTA
              MOVLW    5H
              MOVWF    PORTB
              CALL     CLOCK
              MOVLW    1H
              MOVWF    PORTB
              CALL     CLOCK          ;clock character onto display.
              RETLW    0
```

```
R    MOVLW   2
     MOVWF   PORTA
     MOVLW   5H
     MOVWF   PORTB
     CALL    CLOCK
     MOVLW   2H
     MOVWF   PORTB
     CALL    CLOCK          ;clock character onto display.
     RETLW   0

S    MOVLW   2
     MOVWF   PORTA
     MOVLW   5H
     MOVWF   PORTB
     CALL    CLOCK
     MOVLW   3H
     MOVWF   PORTB
     CALL    CLOCK          ;clock character onto display.
     RETLW   0

T    MOVLW   2
     MOVWF   PORTA
     MOVLW   5H
     MOVWF   PORTB
     CALL    CLOCK
     MOVLW   4H
     MOVWF   PORTB
     CALL    CLOCK          ;clock character onto display.
     RETLW   0

U    MOVLW   2
     MOVWF   PORTA
     MOVLW   5H
     MOVWF   PORTB
     CALL    CLOCK
     MOVLW   5H
     MOVWF   PORTB
     CALL    CLOCK          ;clock character onto display.
     RETLW   0
V    MOVLW   2
     MOVWF   PORTA
     MOVLW   5H
     MOVWF   PORTB
     CALL    CLOCK
```

```
        MOVLW   6H
        MOVWF   PORTB
        CALL    CLOCK       ;clock character onto display.
        RETLW   0

WW      MOVLW   2
        MOVWF   PORTA
        MOVLW   5H
        MOVWF   PORTB
        CALL    CLOCK
        MOVLW   7H
        MOVWF   PORTB
        CALL    CLOCK       ;clock character onto display.
        RETLW   0

X       MOVLW   2
        MOVWF   PORTA
        MOVLW   5H
        MOVWF   PORTB
        CALL    CLOCK
        MOVLW   8H
        MOVWF   PORTB
        CALL    CLOCK       ;clock character onto display.
        RETLW   0

Y       MOVLW   2
        MOVWF   PORTA
        MOVLW   5H
        MOVWF   PORTB
        CALL    CLOCK
        MOVLW   9H
        MOVWF   PORTB
        CALL    CLOCK       ;clock character onto display.
        RETLW   0

Z       MOVLW   2
        MOVWF   PORTA
        MOVLW   5H
        MOVWF   PORTB
        CALL    CLOCK
        MOVLW   0AH
        MOVWF   PORTB
        CALL    CLOCK       ;clock character onto display.
        RETLW   0
```

```
NUM0    MOVLW    2           ;enables the display
        MOVWF    PORTA
        MOVLW    3H
        MOVWF    PORTB
        CALL     CLOCK
        MOVLW    0H
        MOVWF    PORTB
        CALL     CLOCK       ;clock character onto display.
        RETLW    0

NUM1    MOVLW    2           ;enables the display
        MOVWF    PORTA
        MOVLW    3H
        MOVWF    PORTB
        CALL     CLOCK
        MOVLW    1H
        MOVWF    PORTB
        CALL     CLOCK       ;clock character onto display.
        RETLW    0

NUM2    MOVLW    2           ;enables the display
        MOVWF    PORTA
        MOVLW    3H
        MOVWF    PORTB
        CALL     CLOCK
        MOVLW    2H
        MOVWF    PORTB
        CALL     CLOCK       ;clock character onto display.
        RETLW    0

NUM3    MOVLW    2           ;enables the display
        MOVWF    PORTA
        MOVLW    3H
        MOVWF    PORTB
        CALL     CLOCK
        MOVLW    3H
        MOVWF    PORTB
        CALL     CLOCK       ;clock character onto display.
        RETLW    0

NUM4    MOVLW    2           ;enables the display
        MOVWF    PORTA
        MOVLW    3H
        MOVWF    PORTB
```

```
                CALL    CLOCK            ;clock character onto display.
                MOVLW   4H
                MOVWF   PORTB
                CALL    CLOCK            ;clock character onto display.
                RETLW   0

NUM5            MOVLW   2                ;enables the display
                MOVWF   PORTA
                MOVLW   3H
                MOVWF   PORTB
                CALL    CLOCK
                MOVLW   5H
                MOVWF   PORTB
                CALL    CLOCK            ;clock character onto display.
                RETLW   0

NUM6            MOVLW   2                ;enables the display
                MOVWF   PORTA
                MOVLW   3H
                MOVWF   PORTB
                CALL    CLOCK
                MOVLW   6H
                MOVWF   PORTB
                CALL    CLOCK            ;clock character onto display.
                RETLW   0

NUM7            MOVLW   2                ;enables the display
                MOVWF   PORTA
                MOVLW   3H
                MOVWF   PORTB
                CALL    CLOCK
                MOVLW   7H
                MOVWF   PORTB
                CALL    CLOCK            ;clock character onto display.
                RETLW   0

NUM8            MOVLW   2                ;enables the display
                MOVWF   PORTA
                MOVLW   3H
                MOVWF   PORTB
                CALL    CLOCK
                MOVLW   8H
                MOVWF   PORTB
                CALL    CLOCK            ;clock character onto display.
                RETLW   0
```

```
NUM9      MOVLW    2              ;enables the display
          MOVWF    PORTA
          MOVLW    3H
          MOVWF    PORTB
          CALL     CLOCK
          MOVLW    9H
          MOVWF    PORTB
        · CALL     CLOCK          ;clock character onto display.
          RETLW    0


GAP       MOVLW    2
          MOVWF    PORTA
          MOVLW    2H
          MOVWF    PORTB
          CALL     CLOCK
          MOVLW    0H
          MOVWF    PORTB
          CALL     CLOCK          ;clock character onto display.
          RETLW    0



DOT       MOVLW    2
          MOVWF    PORTA
          MOVLW    2H
          MOVWF    PORTB
          CALL     CLOCK
          MOVLW    0EH
          MOVWF    PORTB
          CALL     CLOCK          ;clock character onto display.
          RETLW    0


CLRDISP   CLRF     PORTA
          MOVLW    0H
          MOVWF    PORTB
          CALL     CLOCK          ;clock character onto display.
          MOVLW    1
          MOVWF    PORTB
          CALL     CLOCK
          CALL     DELAYP1
          RETLW    0
```

.***
;

;CONFIGURATION SECTION.

```
START      BSF        STATUS,5        ;Turn to BANK1
           MOVLW      B'00000000'     ;5 bits of PORTA are O/Ps.
           TRIS       PORTA
           MOVLW      B'00000000'
           TRIS       PORTB           ;PORTB IS OUTPUT
           MOVLW      B'00000111'
           OPTION                     ;PRESCALER is /256
           BCF        STATUS,5        ;Return to BANK0
           CLRF       PORTA           ;Clears PORTA
           CLRF       PORTB           ;Clears PORTB
```

;Display Configuration
```
           MOVLW      03H             ;FUNCTION SET
           MOVWF      PORTB           ;8 bit data (default)
           CALL       CLOCK

           CALL       DELAYP1         ;wait for display

           MOVLW      02H             ;FUNCTION SET
           MOVWF      PORTB           ;change to 4 bit
           CALL       CLOCK           ;clock in data

           CALL       DELAYP1         ;wait for display
           MOVLW      02H             ;FUNCTION SET
           MOVWF      PORTB           ;must repeat command
           CALL       CLOCK           ;clock in data

           CALL       DELAYP1         ;wait for display
           MOVLW      08H             ;4 bit micro
           MOVWF      PORTB           ;using 2 line display.
           CALL       CLOCK           ;clock in data

           CALL       DELAYP1
           MOVLW      0H              ;Display on, cursor off
           MOVWF      PORTB           ;0CH
           CALL       CLOCK
           MOVLW      0CH
           MOVWF      PORTB
           CALL       CLOCK
```

```
          CALL     DELAYP1
          MOVLW    0H                    ;Increment cursor, 06H
          MOVWF    PORTB
          CALL     CLOCK
          MOVLW    6H
          MOVWF    PORTB
          CALL     CLOCK
;********************************************************
;
;Program starts now.

BEGIN     CALL     CLRDISP
          CLRF     PORTA
          MOVLW    8H                    ;Cursor at top left, 80H
          MOVWF    PORTB
          CALL     CLOCK
          MOVLW    0H
          MOVWF    PORTB
          CALL     CLOCK

          CALL     M                     ;display M
          CALL     DELAYP1               ;wait 0.1 seconds
          CALL     I                     ;display I
          CALL     DELAYP1               ;wait 0.1 seconds
          CALL     C                     ;Etc.
          CALL     DELAYP1
          CALL     R
          CALL     DELAYP1
          CALL     O
          CALL     DELAYP1
          CALL     C
          CALL     DELAYP1
          CALL     O
          CALL     DELAYP1
          CALL     N
          CALL     DELAYP1
          CALL     T
          CALL     DELAYP1
          CALL     R
          CALL     DELAYP1
          CALL     O
          CALL     DELAYP1
          CALL     L
          CALL     DELAYP1
          CALL     L
```

```
        CALL    DELAYP1
        CALL    E
        CALL    DELAYP1
        CALL    R
        CALL    DELAYP1
        CALL    S
        CALL    DELAYP1

        CLRF    PORTA
        MOVLW   0CH              ;Cursor on second line, C3
        MOVWF   PORTB
        CALL    CLOCK
        MOVLW   3H
        MOVWF   PORTB
        CALL    CLOCK

        CALL    A
        CALL    DELAYP1
        CALL    T
        CALL    DELAYP1
        CALL    GAP
        CALL    T
        CALL    DELAYP1
        CALL    H
        CALL    DELAYP1
        CALL    E
        CALL    DELAYP1
        CALL    GAP
        CALL    M
        CALL    DELAYP1
        CALL    M
        CALL    DELAYP1
        CALL    U
        CALL    DELAYP1


        CALL    DELAY3           ;wait 3 seconds

        CALL    CLRDISP

        MOVLW   8H               ;Cursor at top left, 80H
        MOVWF   PORTB
        CALL    CLOCK
        MOVLW   0H
```

```
MOVWF    PORTB
CALL     CLOCK

CALL     C
CALL     DELAYP1
CALL     O
CALL     DELAYP1
CALL     N
CALL     DELAYP1
CALL     T
CALL     DELAYP1
CALL     A
CALL     DELAYP1
CALL     C
CALL     DELAYP1
CALL     T
CALL     DELAYP1

CLRF     PORTA
MOVLW    0CH              ;Cursor on 2nd line
MOVWF    PORTB
CALL     CLOCK
MOVLW    3H
MOVWF    PORTB
CALL     CLOCK

CALL     D
CALL     DELAYP1
CALL     A
CALL     DELAYP1
CALL     V
CALL     DELAYP1
CALL     E
CALL     DELAYP1
CALL     GAP
CALL     DELAYP1
CALL     S
CALL     DELAYP1
CALL     M
CALL     DELAYP1
CALL     I
CALL     DELAYP1
CALL     T
CALL     DELAYP1
```

```
        CALL    H

        CALL    DELAY3          ;wait 3 seconds

        GOTO    BEGIN
END
```

Program operation

• PORTA and PORTB are configured as outputs in the CONFIGURATION SECTION.

Display configuration

• In the Display Configuration Section, the Register Select (R/S) line, A1 on the microcontroller, is set low by CLRF PORTA in the Configuration Section.
• R/S = 0 ensures that the data to the display will change the registers. Later R/S = 1 writes the characters to the display.
• The display is expecting its data to arrive via eight lines, but to save I/O lines we will use four and write them twice. The code to do this and also tell the driver chip the display is a two line display is:

```
        MOVLW   03H             ;FUNCTION SET
        MOVWF   PORTB           ;8 bit data (default)
        CALL    CLOCK

        CALL    DELAYP1         ;wait for display

        MOVLW   02H             ;FUNCTION SET
        MOVWF   PORTB           ;change to 4 bit
        CALL    CLOCK           ;clock in data

        CALL    DELAYP1         ;wait for display
        MOVLW   02H             ;FUNCTION SET
        MOVWF   PORTB           ;must repeat command
        CALL    CLOCK           ;clock in data

        CALL    DELAYP1         ;wait for display
        MOVLW   08H             ;4 bit micro
        MOVWF   PORTB           ;using 2 line display.
        CALL    CLOCK           ;clock in data
```

The data is set up on PORTB using B0, 1, 2 and 3. As in

```
MOVLW    03H               ;FUNCTION SET
MOVWF    PORTB
```

This data is then clocked into the display by pulsing the Enable line (E, A2 on the micro) high and then low with:

```
CLOCK    BSF      PORTA,2
         NOP
         BCF      PORTA,2
         NOP
         RETLW    0
```

CALL DELAYP1, waits for 0.1 seconds to give the display time to activate before continuing.

When the display has been configured to turn on, switch the cursor off, and increment the cursor after every character write, we are then ready to write to the display.

Writing to the display

- The display is cleared if required with:

```
CALL CLRDISP
```

- The address of the character is first written to the display, say the 80H position (top left-hand corner).

```
CLRF     PORTA
MOVLW    8H               ;Cursor at top left, 80H
MOVWF    PORTB
CALL     CLOCK
MOVLW    0H
MOVWF    PORTB
CALL     CLOCK
```

Notice the 8 is sent first followed by the 0. To write to the position mid-way along the top line the address would be 88H. So the 80H in the code above would be replaced by 88H.

- In order to write the letter 'M' in the display at the position defined, we CALL M and use the code 4DH, NB. Send the 4 first followed by the D. The Register Select (R/S) line A1 on the micro is set to 1 for the character write option. The code is:

```
M       MOVLW   2                ;enables the display
        MOVWF   PORTA            ;sets A1=1
        MOVLW   4H               ;send data 4
        MOVWF   PORTB
        CALL    CLOCK
        MOVLW   0DH              ;send data D
        MOVWF   PORTB
        CALL    CLOCK            ;clock character 'M' onto display.
        RETLW   0
```

In this way any one of the 192 characters available can be shown on the display.

The program continues by printing out the rest of the message. A delay of 0.1 seconds is maintained after printing each character to give the effect of the message being typed out.

All the capital letters and numbers 0 to 9 have been included in the header so you can easily enter your own message. The complete character set for the display showing all 192 characters is illustrated in Figure 7.4.

Displaying a number

Suppose we wish to display a number thrown by a dice, for example a 4. We could use the instruction CALL NUM4, but we would not have known previously that the number was going to be a 4. The throw of the dice would be stored in a user file called, say, THROW and this would then have 4 in it.

Now the code for 0 is 30H
The code for 1 is 31H
The code for 2 is 32H
Etc.

If we wanted to display the number 4 the code is:

```
NUM4    MOVLW   2                ;enables the display
        MOVWF   PORTA
        MOVLW   3H               ;34H is the code for 4
        MOVWF   PORTB
        CALL    CLOCK
```

```
MOVLW     4H
MOVWF     PORTB
CALL      CLOCK           ;clock character onto display.
RETLW     0
```

If the 4 is in the file THROW, we can display this with the code:

```
MOVLW     2               ;enables the display
MOVWF     PORTA
MOVLW     3H
MOVWF     PORTB
CALL      CLOCK
MOVF      THROW,W         ;number comes from the file
MOVWF     PORTB
CALL      CLOCK           ;clock character onto display.
RETLW     0
```

Notice how the value of the number now has come from the file.
This code would then display any number in the file THROW.

If you measured a temperature as 27°C, you would probably store the 2 in a file TEMPTENS (tens of degrees) and the 7 in a file TEMPUNIT (units of degrees).

You can then modify the code above to display:

THE TEMPERATURE
IS 27°C.

The 'I' would be located at address C5H on the display. The temperature would then be written at locations C8H and C9H. There would be no need to rewrite the message, just rewrite the temperature as it changed, after first moving the cursor to address C8H.

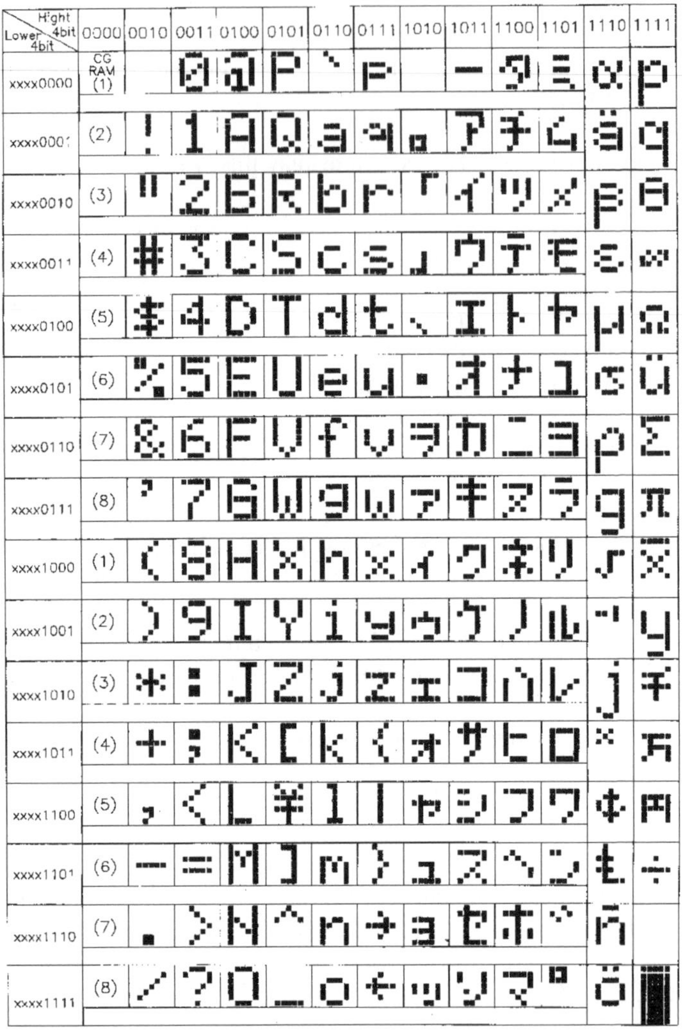

Figure 7.4 Alphanumeric display character set

8
Analogue to digital conversion

Up to now we have considered inputs as being digital in operation, i.e. the input is either a 0 or 1. But suppose we wish to make temperature measurements, but not just hot or cold (1 or 0). We may, for example, require to:

- Sound a buzzer if the temperature drops below freezing.
- Turn a heater on if the temperature is below 18°C.
- Turn on a fan if the temperature goes above 25°C.
- Turn on an alarm if the temperature goes above 30°C.

We could of course have separate digital inputs, coming from comparator circuits for each setting. But a better solution is to use one input connected to an analogue to digital converter and measure the temperature with that.

Figure 8.1 shows a basic circuit for measuring temperature. It consists of a fixed resistor in series with a thermistor (a temperature sensitive resistor).

The resistance of the thermistor changes with temperature causing a change in the voltage at point X in Figure 8.1. As the temperature rises the voltage at X rises. As the temperature decreases the voltage at X reduces.

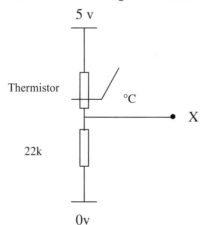

Figure 8.1 Temperature measuring circuit

We need to know the relationship between the temperature of the thermistor and the voltage at X. A simple way of doing this would be to place the thermistor in a cup of boiling water (100°C) and measure the voltage at X. As the water cools corresponding readings of temperature and voltage can be taken. If needed a graph of these temperature and voltage readings could be plotted.

Making an A/D reading

In the initial example let us suppose:

- 0°C gave a voltage reading of 0.6v
- 18°C gave a reading of 1.4v
- 25°C gave a reading of 2.4v
- 30°C gave a reading of 3.6v

The microcontroller would read these voltages and convert them to an 8-bit number where 0v is 0 and 5v is 255. i.e. a reading of 51 per volt or a resolution of 1/51v, i.e. 1 bit is 19.6mv.

So 0°C = 0.6v = reading of 31 (0.6 × 51 = 30.6)
18°C = 1.4v = 71 (1.4 × 51 = 71.4)
25°C = 2.4v = 122 (2.4 × 51 = 122.4)
30°C = 3.6v = 184 (3.6 × 51 = 183.6)

If we want to know when the temperature is above 30°C the microcontroller looks to see if the A/D reading is above 184. If it is, switch on the alarm, if not keep the alarm off. In a similar way any other temperature can be investigated – not just the ones listed. With our 8 bits we have 255 different temperatures we can choose from. The PIC16C773 and PIC16C774 have 12-bit A/D converters and can have 4096 different temperature points.

Analogue to digital conversion was introduced to the PIC microcontrollers with the family 16C7X devices: 16C71, 16C73 and 16C74. Table 8.1 shows some of the specifications of these devices.

Table 8.1 16C7X device specifications

Device	I/O	A/D Channels	Program Memory	Data Memory	Current Source/Sink
16C71	13	4	1k	36	25mA
16C73	22	5	4k	192	25mA
16C74	33	8	4k	192	25mA

This family of devices has now been extended to include others called 16C71X. Table 8.2 shows these additions.

Table 8.2 16C71X device additions

Device	I/O	A/D Channels	Program Memory	Data Memory	Current Source/Sink
16C710	13	4	512	36	25mA
16C711	13	4	1k	68	25mA
16C715	13	4	2k	128	25mA
16C72	22	5	2k	128	25mA
16C76	22	5	8k	368	25mA
16C77	33	8	8k	368	25mA

There are also a number of other chips containing A/D converters such as the 16F873, 16F874, 16F876 and 16F877. They are also flash devices and do not need erasing with UV light.

The basic device in Tables 8.1 and 8.2 is the 16C71, but this is now being replaced with the 16C711. In this section we will consider the 16C711 device but all the programs are transferable to the others. The 16C711 device is not a flash device, so for development work a 16C711JW is required. The 16C711 has four A/D inputs (channels) and can make four simultaneous analogue measurements.

The 16C711 device needs extra registers, which the 16F84 does not have, to handle the A/D processing. The memory map of the 16C711 including these registers is shown in Figure 8.2.

FILE ADDRESS	FILE NAME	FILE NAME
00	INDIRECT ADDRESS	INDIRECT ADDRESS
01	TMR0	OPTION
02	PCL	PCL
03	STATUS	STATUS
04	FSR	FSR
05	PORTA	TRISA
06	PORTB	TRISB
07	—	—
08	ADCON0	ADCON1
09	ADRES	ADRES
0A	PCLATH	PCLATH
0B	INTCON	INTCON
0C	68	
	USER	
4F	FILES	
	BANK0	BANK1

Figure 8.2 Memory map of the 16C711

The 16C711 has four analogue inputs AN0, AN1, AN2 and AN3. But these pins are also used as A0, A1, A2 and A3/Vref (digital pins and voltage reference). See Appendix A for the pinout of the 16C711.

Configuring the A/D device

In order to make an analogue measurement we have to configure the device. To ease the programming a header has been done for the 16C711 in a file HEADER71.ASM. The header sets up PORTA as an input port with pins 17 and 18, AN0 and AN1 are set as analogue inputs. Pins 1 and 2, A2 and A3 have been set up as digital inputs and the voltage reference has been set at 5v, i.e. Vdd. PORTB has been set as an output port.

To configure the 16C711 for A – D measurements three registers need to be set up: ADCON0, ADCON1 and ADRES.

ADCON0

The first of the A/D registers, ADCON0, is A to D Control Register 0.
ADCON0 is used to:

- Switch the A/D converter on with ADON, bit 0. This bit turns the A/D on when set and off when clear. Once it is turned on the A/D can be left on all of the time but it does draw a current of 90μA, compared to the rest of the microcontroller which draws a current of 15μA.
- Instruct the microcontroller to execute a conversion by setting the GO/DONE bit, bit 2. When the GO/DONE bit is set the micro does an A/D conversion. When the conversion is complete the hardware clears the GO/DONE bit. This bit can be read to determine when the result is ready.
- Set the particular channel (input) to make the measurement from. This is done with two Channel Select bits, CHS0 and CHS1, bits 3 and 4.

The register ADCON0 is shown in Figure 8.3.

ADCON1

In ADCON1, A to D Conversion register 1, only bits 0 and 1 are used.
They are the Port Configuration bits, PCFG0 and PCFG1, that determine which of the pins on PORTA will be analogue inputs and which will be digital.

The ADCON1 register is illustrated in Figure 8.4 and the corresponding analogue and digital inputs are shown in Figure 8.5.

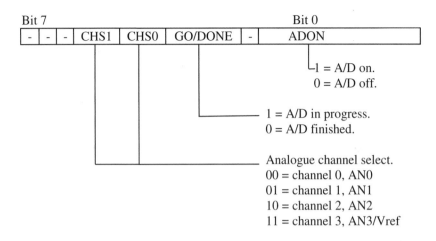

Figure 8.3 ADCON0 register

Figure 8.4 ADCON1 register

PCFG	A0	A1	A2	A3	Ref
00	A	A	A	A	Vdd
01	A	A	A	Vref	A3
10	A	A	D	D	Vdd
11	D	D	D	D	Vdd

Figure 8.5 ADCON1 Port configuration

To provide a mixture of analogue and digital inputs I have configured the 16C711 header with ADCON1 = 10, i.e. B'00000010', to give two analogue and two digital inputs.

ADRES

- The third register is ADRES, the A to D RESult register. This is the file where the result of the A/D conversion is stored. If several measurements require storing then the number in ADRES needs to be transferred to a user file before it is overwritten with the next measurement.

Header for the 16C711

```
;HEADER71.ASM for 16C711.   This sets PORTA as an INPUT with A0 and A1 as
;                           analogue inputs and A2 and A3 as digital.
;                           PORTB is an OUTPUT. The OPTION Register is
;                           set to /256 to give timing pulses of 1/32 of a second
;                           1 second and 0.5 second delays are included in the
;                           subroutine section.
;                           The A/D converter is turned on.
;******************************************************
```

;EQUATES SECTION

```
TMR0        EQU     1        ;means TMR0 is file 1.
STATUS      EQU     3        ;means STATUS is file 3.
PORTA       EQU     5        ;means PORTA is file 5.
PORTB       EQU     6        ;means PORTB is file 6.
ZEROBIT     EQU     2        ;means ZEROBIT is bit 2.
CARRY       EQU     0
ADCON0      EQU     8        ;A/D Configuration reg.0
ADCON1      EQU     8        ;A/D Configuration reg.1
ADRES       EQU     9        ;A/D Result register.
COUNT       EQU     0CH      ;means COUNT is file 0C,
                            ;a register to count events.
;******************************************************

LIST        P=16C711     ;we are using the 16C711.
ORG         0            ;the start address in memory is 0
GOTO        START        ;goto start!

;******************************************************
```

;SUBROUTINE SECTION.

```
;1 second delay.
DELAY1      CLRF    TMR0             ;Start TMR0.
LOOPA       MOVF    TMR0,W           ;Read TMR0 INTO W.
            SUBLW   .32              ;TIME - 32
            BTFSS   STATUS,ZEROBIT   ;Check TIME-W = 0
            GOTO    LOOPA            ;Time is not = 32.
            RETLW   0                ;Time is 32, return.

;0.5 second delay.
DELAYP5     CLRF    TMR0             ;Start TMR0.
```

```
LOOPB      MOVF       TMR0,W              ;Read TMR0 INTO W.
           SUBLW      .16                 ;TIME - 16
           BTFSS      STATUS,ZEROBIT      ;Check TIME-W = 0
           GOTO       LOOPB               ;Time is not = 16.
           RETLW      0                   ;Time is 16, return.
```

;**

;CONFIGURATION SECTION

```
START      BSF        STATUS,5            ;Turns to Bank1.
           MOVLW      B'00011111'         ;5bits of PORTA are I/P
           TRIS       PORTA
           MOVLW      B'00000010'         ;A0, A1 are analogue
           MOVWF      ADCON1              ;A2, A3 are digital I/P.
           MOVLW      B'00000000'
           TRIS       PORTB               ;PORTB is OUTPUT
           MOVLW      B'00000111'         ;Prescaler is /256
           OPTION                         ;TIMER is 1/32 secs.
           BCF        STATUS,5            ;Return to Bank0.
           MOVLW      B'00000001'         ;Turn on A/D converter.
           MOVWF      ADCON0              ;and selects channel A0
           CLRF       PORTA               ;Clears PortA.
           CLRF       PORTB               ;Clears PortB.
```

;**
;Program starts now.

HEADER71.ASM explained

HEADER71.ASM is similar in operation to Header84.ASM outlined in Chapter 3, with the following exceptions:

- In the Equates Section, ADCON0 is equated to file 8 and ADCON1 is equated to file 8! The reason that they can share the same file number is that ADCON1 is in Bank1 and ADCON0 is in Bank0. So that the page (Bank) select bit needs to be set correctly in the status register. The Carry Bit in the status register, which indicates if a calculation is +ve or −ve, is bit 0 and has been equated to 0.
- The software is told we are using the 16C711 with LIST P = 16C711.
- In the Configuration Section A0 and A1 are set as analogue inputs, A2 and A3 are set up as digital inputs with:

```
MOVLW      B'00000010'
```

```
MOVWF      ADCON1
```

The A/D converter is switched on with:

```
MOVLW      B'00000001'
MOVWF      ADCON0
```

A/D conversion – example, a temperature sensitive switch

To introduce the working of the A/D converter we will consider a simple example, i.e. turn an LED on when the temperature is above 25°C and turn the LED off when it is below 25°C.

The diagram for this temperature switch circuit is shown in Figure 8.6.

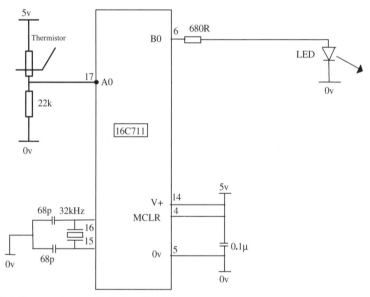

Figure 8.6 Temperature switch circuit

Taking the A/D reading

The A/D converter has been switched on in the header and it automatically looks at Channel 0 unless told otherwise. In order to make the measurement the GO/DONE bit, bit 2 is set and we wait until it is cleared with:

```
           BSF     ADCON0,2      ;Take measurement, set GO/DONE
WAIT       BTFSC   ADCON0,2      ;Wait until GO/DONE is clear
           GOTO    WAIT
```

The measurement will then be in the A/D Result register, ADRES.

Determining if the temperature is above or below 25°C

Suppose the voltage on the analogue input, Channel 0, A0 is 2.4v when the temperature is 25°C. The required A/D reading for 2.4v is $2.4 \times 51 = 122$. We therefore need to know when the A/D reading is above and below 122, i.e. above and below 25°C.

Previously we have seen how to tell if a value is equal to another by subtracting and looking at the zerobit in the status register (Chapter 3). There is another bit, bit 0, in the status register called the Carry Bit, which indicates if the result of a subtraction is +ve or –ve. If the Carry Bit is set the result was +ve, if the bit is clear the result was –ve. So we can tell if the number is above or below a defined value. The code for this is:

```
MOVF     ADRES,W        ;Move Analogue result into W
SUBLW    .122           ;Do 122 – ADRES, i.e. 122 - W
BTFSC    Status,Carry   ;Check the carry bit. Clear if ADRES>122
GOTO     TURNOFF        ;Routine to turn off LED
GOTO     TURNON         ;Routine to turn on LED
```

The analogue measurement is moved from ADRES into W where we can subtract it from 122. Note: The subtraction always does, Value – W. The carry bit tells us if the A/D result is above or below 122. Note: If the result of the subtraction is zero the carry is also 1. It must be 1 or 0. Being +ve or zero does not matter in this example.

We have then found out if the result is equal to or above 122, or if it is less than 122.

When the measurement is made we then goto one of two subroutines, TURNON or TURNOFF. These subroutines are not very grand but they could easily be more complicated, sometimes hundreds of lines long.

Program code

The full code for this temperature sensitive switch program is shown below as TEMPSENS.ASM.

```
;TEMPSENS.ASM.   This sets PORTA as an INPUT with A0
;                and A1 as analogue inputs and A2 and A3
;                as digital. PORTB is an OUTPUT.
;*********************************************************
```

;EQUATES SECTION

```
TMR0        EQU    1          ;means TMR0 is file 1.
STATUS      EQU    3          ;means STATUS is file 3.
PORTA       EQU    5          ;means PORTA is file 5.
PORTB       EQU    6          ;means PORTB is file 6.
CARRY       EQU    0
ADCON0      EQU    8
ADCON1      EQU    8
ADRES       EQU    9
```

;CONFIGURATION SECTION

```
START       BSF    STATUS,5         ;Turns to Bank1.
            MOVLW  B'00011111'      ;5 bits of PORTA are I/P
            TRIS   PORTA
            MOVLW  B'00000010'      ;A0, A1 are analogue
            MOVWF  ADCON1           ;A2, A3 are digital I/P.
            MOVLW  B'00000000'
            TRIS   PORTB            ;PORTB is OUTPUT
            BCF    STATUS,5         ;Return to Bank0.
            MOVLW  B'00000001'      ;Turns on A/D converter,
            MOVWF  ADCON0           ;and selects channel AN0
            CLRF   PORTA            ;Clears PortA.
            CLRF   PORTB            ;Clears PortB.
```

;**
;Program starts now.

```
BEGIN       BSF    ADCON0,2         ;Take measurement, set GO/DONE
WAIT        BTFSC  ADCON0,2         ;Wait until GO/DONE is clear
            GOTO   WAIT

            MOVF   ADRES,W          ;Move Analogue result into W
            SUBLW  .122             ;Do 122 – ADRES, i.e. 122 - W
            BTFSC  STATUS,CARRY     ;Clear if ADRES>122
            GOTO   TURNOFF          ;Routine to turn off LED
            GOTO   TURNON           ;Routine to turn on LED
TURNOFF     BCV    PORTB, 0
            GOTO   BEGIN

TURNON      BSF    PORTB, 0
            GOTO   BEGIN
END
```

Another example – a voltage indicator

Previously we have looked at a single input level. But with our 8-bit micro we could look at 255 different input levels.

Suppose we wish to use the LEDs connected to PORTB to indicate the voltage on the analogue input AN0. So that as the voltage increases the number of LEDs lit also increases.

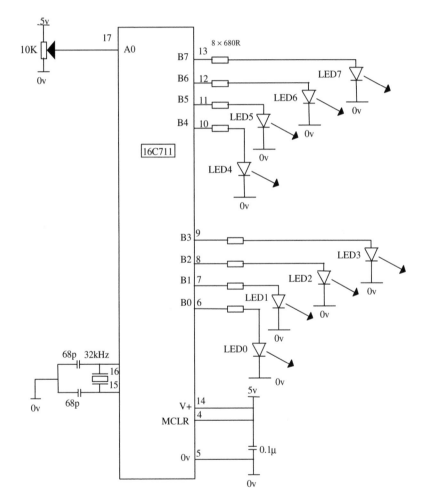

Figure 8.7 Circuit for the voltage indicator

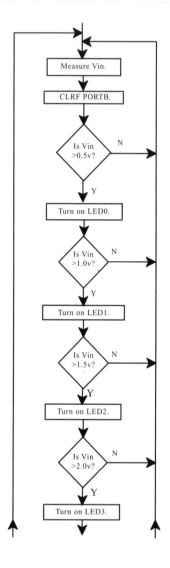

Figure 8.8 Flowchart for the voltage indicator

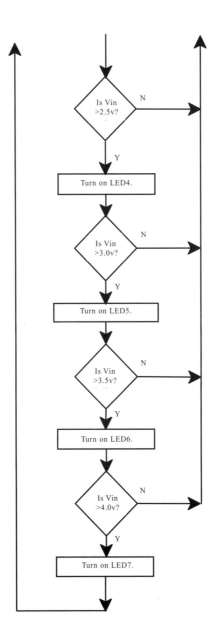

In HEADER71.ASM we have configured the micro so that the voltage reference is Vdd, i.e. the 5v supply. This was done with the instructions:

```
MOVLW   B'00000010'
MOVWF   ADCON1
```

This means that 5v will give a digital reading of 255 in our 8-bit register ADRES. The resolution of this register is 5v/255 = 19.6mV.

Suppose our LED ladder was to increment in 0.5v steps as indicated below:

Vin = 0 - 0.5v	All LEDs off,	$0.5v = 0.5/5 \times 255 = 25.5 = 26$
Vin = 0.5 - 1.0v	B0 on,	$1.0v = 1/5 \times 255 = 51$
Vin = 1.0 - 1.5v	B1 on,	$1.5v = 1.5/5 \times 255 = 76.5 = 77$
Vin = 1.5 - 2.0v	B2 on,	$2.0v = 2/5 \times 255 = 102$
Vin = 2.0 - 2.5v	B3 on,	$2.5v = 2.5/5 \times 255 = 127.5 = 128$
Vin = 2.5 - 3.0v	B4 on,	$3.0v = 3/5 \times 255 = 153$
Vin = 3.0 - 3.5v	B5 on,	$3.5v = 3.5/5 \times 255 = 178.5 = 179$
Vin = 3.5 - 4.0v	B6 on,	$4.0v = 4/5 \times 255 = 204$
Vin = 4.0 - 5.0v	B7 on.	

The circuit diagram for this voltage indicator is shown in Figure 8.7 and the flowchart is shown in Figure 8.8.

Voltage indicator, program solution

HEADER71.ASM is altered to produce the program VOLTIND.ASM for the voltage indicator circuit.

```
; VOLTIND.ASM.   This sets PORTA as an INPUT with A0
;                and A1 as analogue inputs and A2 and A3
;                as digital. PORTB is an OUTPUT.
;**********************************************************
```

;EQUATES SECTION

TMR0	EQU	1	;means TMR0 is file 1.
STATUS	EQU	3	;means STATUS is file 3.
PORTA	EQU	5	;means PORTA is file 5.
PORTB	EQU	6	;means PORTB is file 6.
ZEROBIT	EQU	2	;means ZEROBIT is bit 2.
ADCON0	EQU	8	;A/D Configuration reg.0
ADCON1	EQU	8	;A/D Configuration reg.1

```
ADRES       EQU     9           ;A/D Result register.
CARRY       EQU     0           ;Carry bit in Status Register
COUNT       EQU     0CH         ;means COUNT is file 0C,
                                ;a register to count events.
;********************************************************

LIST        P=16C711    ;we are using the 16C711.
ORG         0           ;the start address in memory is 0
GOTO        START       ;goto start!

;********************************************************
;CONFIGURATION SECTION

START       BSF     STATUS,5    ;Turns to Bank1.
            MOVLW   B'00011111' ;5bits of PORTA are I/P
            TRIS    PORTA
            MOVLW   B'00000010' ;A0, A1 are analogue
            MOVWF   ADCON1      ;A2, A3 are digital I/P.
            MOVLW   B'00000000'
            TRIS    PORTB       ;PORTB is OUTPUT
            BCF     STATUS,5    ;Return to Bank0.
            MOVLW   B'00000001' ;Turns on A/D converter,
            MOVWF   ADCON0      ;and selects channel AN0
            CLRF    PORTA       ;Clears PortA.
            CLRF    PORTB       ;Clears PortB.

;********************************************************
;Program starts now.

BEGIN       BSF     ADCON0,2    ;Take Measurement.
WAIT        BTFSC   ADCON0,2    ;Wait until reading done.
            GOTO    WAIT
            MOVF    ADRES,W     ;Move A/D Result into W
            CLRF    PORTB       ;Clear PortB.

            SUBLW   .26         ;26-,W. W is altered
            BTFSC   STATUS,CARRY ;Is W> or < 26
            GOTO    BEGIN       ;W is < 26 (0.5v)

            MOVF    ADRES,W     ;Move A/D Result into W
            BSF     PORTB,0     ;Turn on B0.
            SUBLW   .51         ;51-,W. W is altered
            BTFSC   STATUS,CARRY ;Is W> or < 51
            GOTO    BEGIN       ;W is < 51 (1.0v)
```

```
MOVF        ADRES,W         ;Move A/D Result into W
BSF         PORTB,1         ;Turn on B1.
SUBLW       .77             ;77-,W. W is altered
BTFSC       STATUS,CARRY    ;Is W> or < 77
GOTO        BEGIN           ;W is < 77 (1.5v)

MOVF        ADRES,W         ;Move A/D Result into W
BSF         PORTB,2         ;Turn on B2.
SUBLW       .102            ;102-,W. W is altered
BTFSC       STATUS,CARRY    ;Is W> or < 102
GOTO        BEGIN           ;W is < 102 (2.0v)

MOVF        ADRES,W         ;Move A/D Result into W
BSF         PORTB,3         ;Turn on B3.
SUBLW       .128            ;128-,W. W is altered
BTFSC       STATUS,CARRY    ;Is W> or < 128
GOTO        BEGIN           ;W is < 128 (2.5v)

MOVF        ADRES,W         ;Move A/D Result into W
BSF         PORTB,4         ;Turn on B4.
SUBLW       .153            ;153 -,W. W is altered
BTFSC       STATUS,CARRY    ;Is W> or < 153
GOTO        BEGIN           ;W is < 153 (3.0v)

MOVF        ADRES,W         ;Move A/D Result into W
BSF         PORTB,5         ;Turn on B5.
SUBLW       .179            ;179 -,W. W is altered
BTFSC       STATUS,CARRY    ;Is W> or < 179
GOTO        BEGIN           ;W is < 179 (3.5v)

MOVF        ADRES,W         ;Move A/D Result into W
BSF         PORTB,6         ;Turn on B6.
SUBLW       .204            ;204 -,W. W is altered
BTFSC       STATUS,CARRY    ;Is W> or < 204
GOTO        BEGIN           ;W is < 204 (4.0v)

BSF         PORTB,7         ;Turn on B7.
GOTO        BEGIN
END
```

Operation of the voltage indicator program

The code to make the analogue measurement is the same as in the temperature switch circuit. Once the measurement has been taken the program checks to see if the digital value of the input is >26, then if it is B0 LED is switched on. The program then checks to see if the measurement is >51, if so then B1 LED is lit. If the reading is >77 then B2 LED is lit, etc. When the value is less than the one being checked then the program branches back to the beginning, makes another measurement and the cycle repeats. Note: After the A/D reading the LEDs are cleared before being turned on, in case the voltage has dropped.

To check if a reading (or any number) is > say 26:

Put the number into W.
Take W from 26, i.e. 26 – W by SUBLW .26.

If the result is +ve, the number is <26 and the carry bit is set in the status register. If the number is >26 the result is –ve and the carry bit is clear.

Problem

To check your understanding of the previous section, try this. Turn a red LED on only when the input voltage is above 3v and turn a yellow LED on only when the input voltage is below 1v and turn a green LED on only when the voltage is between 1v and 3v.

9
Radio transmitters and receivers

I used to find radio circuits a bit daunting but now with the introduction of low cost modules the radio novice like myself can transmit data easily.

This section details the use of the 418MHz radio transmitter and receiver modules (RT1-418 and RR3-418). They do not need a licence to operate and there are many varieties available. The transmitters only have three connections, two power supply and one data input, the transmitting aerial is incorporated on the unit. The receiver has four connections: two power supply, one aerial input and one data output. The receiving aerial only needs to be a piece of wire about 25cm long.

Radio data transmission system

The basic circuit diagram of the radio system is shown in Figure 9.1.

The microcontroller generates the data and then passes the data pulses to the transmitter. The receiver receives the data pulses and a microcontroller decodes the information and processes it.

A microcontroller system could measure the temperature outside and transmit this temperature to be displayed on a unit inside.

How does it work?

The transmitter

Data is generated by the microcontroller, say by pressing a switch or from a temperature sensor via the 16C711 doing an A/D conversion. Suppose this data

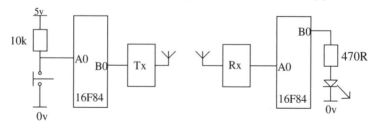

Figure 9.1 Radio data transmission system

is 27H, this would then be stored in a user file called, say, NUMA. So file NUMA would appear as shown in Figure 9.2.

NUMA,7	NUMA,6	NUMA,5	NUMA,4	NUMA,3	NUMA,2	NUMA,1	NUMA,0
0	0	1	0	0	1	1	1

Figure 9.2 File NUMA containing 27H

The data then needs to be passed from the micro to the data input of the transmitter. The transmitter output will then be turned on and off by the data pulses. The length of time the transmitter is on will indicate if the data was a 1, a 0 or the transmission start pulse.

I have decided to use a start bit that is 7.5ms wide, a 5ms pulse to represent a logic 1 and a 2.5ms pulse to represent a logic 0. All pulses are separated by a space of 2.5ms. The pulse train for NUMA is then as shown in Figure 9.3.

Pulse train

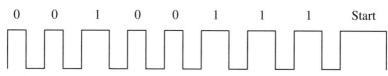

Figure 9.3 NUMA pulse train

In order to generate this train the software turns the output on for the 7.5ms start pulse, off for 2.5ms, on for 5ms for the first 1, off for 2.5ms, on for 5ms for the next logic 1, off for 2.5ms, on for 5ms for the next logic 1, off for 2.5ms, on for 2.5ms for the logic 0, etc.

To generate the data each bit in the file NUMA is tested in turn. If the bit is 0 then the output is turned on for 2.5ms, if the bit is 1 then the output is turned on for 5ms. The code for this data would be:

```
        BSF     PORTB,0     ;Transmit start pulse
        CALL    DELAY3      ;7.5ms Start pulse
        BCF     PORTB,0     ;Transmit space
        CALL    DELAY1      :Delay 2.5ms

TESTA0  BTFSC   NUMA,0      ;Test NUMA,0
        GOTO    SETA0       ;If NUMA0 = 1
        GOTO    CLRA0       ;If NUMA0 = 0

SETA0   BSF     PORTB,0     ;Transmit 1
        CALL    DELAY2      ;Delay 5ms
```

```
              GOTO      TESTA1
CLRA0         BSF       PORTB,0        ;Transmit 0
              CALL      DELAY1         ;Delay 2.5ms
              GOTO      TESTA1


TEASTA1       BCF       PORTB,0        ;Transmit space
              CALL      DELAY1
              BTFSC     NUMA,1         ;Test NUMA,1
              GOTO      SETA1          ;If NUMA0 = 1
              GOTO      CLRA1          ;If NUMA0 = 0


SETA1         BSF       PORTB,0
              CALL      DELAY2
              GOTO      TESTA2

CLRA1         BSF       PORTB,0
              CALL      DELAY1
              GOTO      TESTA2
```

This bit testing is repeated until all 8 bits are transmitted.

The receiver

The receiver works the opposite way round. The data is received and stored in a file NUMA. Several data bytes could be transmitted depending on how many switches are used. Or the data may be continually varying from a temperature sensor. In this example we are only looking for 1 byte, i.e. the number 27H which was transmitted.

The data is passed from the receiver to the input A0 of the microcontroller.

We wait to receive the 7.5ms start bit. When this is detected we then measure the next eight pulses.

If a pulse is 5ms wide then a 1 has been transmitted and we SET the relative bit in the file NUMA. If the pulse is only 2.5ms long then we leave the bit CLEAR.

Measuring the received pulse width

Measuring the width of a pulse is a little more difficult than setting a pulse width. Consider the pulse in Figure 9.4.

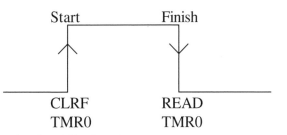

Figure 9.4 Measuring the width of a pulse

The input is continually tested until it goes high and then the timer, TMR0, is cleared to start timing. The input is continually tested until it goes low and then the value of TMR0 is read. This is done by:

> MOVF TMR0,W which puts the value of TMR0 into W.

We can then check to see if the pulse is 5ms long, i.e. a logic 1, if not then a shorter pulse means a logic 0 was transmitted. If the pulse is greater than 3.5ms then it must be a logic1, at 5ms. If the pulse is less than 3.5ms then it must be a logic0. TMR0 will hold a value of 3 after a time of 3.5ms, so we check to see if the width of the pulse is greater or less than 3.

The code for this is:

```
TESTA0H    BTFSS    PORTA,0          ;wait for Hi transmission
           GOTO     TESTA0H
           CLRF     TMR0             ;start timing
TESTA0L    BTFSC    PORTA,0          ;wait for Lo transmission
           GOTO     TESTA0L
           MOVF     TMR0,W           ;read value of TMR0
           SUBLW    3                ;3-W or 3-TMR0
           BTFSC    STATUS,CARRY     ;Is TMR0 3 i.e. a logic1
           BSF      NUMA,0           ;Yes.
```

This measuring of the pulse width continues until all eight pulses are read and the relevant bits stored in the file NUMA. A TMR0 value >6 indicates the pulse was a Start pulse.

We then check to see if the number stored in the file NUMA is 27H. This is done as we have done before by subtracting 27H from it, if the answer is zero, i.e. 27 − 27 = 0, then the number transmitted was 27H and we turn on the LED. It seems such a waste to go to all this trouble to turn an LED on. I hope you can be a little more imaginative – this is only an example.

Radio transmitter/receiver program

The complete codes for the transmitter and receiver are shown below as
TX.ASM and RX.ASM. The OPTION register has been set to produce timing
pulses of 1ms.

Transmitter program code

TX.ASM
;tx.asm transmits code from a switch.

```
TMR0        EQU      1           ;TMR0 is FILE 1.
PORTA       EQU      5           ;PORTA is FILE 5.
PORTB       EQU      6           ;PORTB is FILE 6.
STATUS      EQU      3           ;STATUS is FILE3.
ZEROBIT     EQU      2           ;ZEROBIT is Bit 2.
NUMA        EQU      0DH
;********************************************************
LIST        P=16F84              ;We are using the 16F84.
ORG         0                    ;0 is the start address.
GOTO        START                ;goto start!


;********************************************************
;SUBROUTINE SECTION.

;2.5ms SECOND DELAY
DELAY1      CLRF     TMR0             ;Start TMR0
LOOPA       MOVF     TMR0,W           ;Read TMR0 into W
            SUBLW    .1               ;TIME – W
            BTFSS    STATUS,ZEROBIT   ;Check TIME-W=0
            GOTO     LOOPA
            RETLW    0                ;Return after TMR0 = 32

;5ms SECOND DELAY
DELAY2      CLRF     TMR0             ;Start TMR0
LOOPB       MOVF     TMR0,W           ;Read TMR0 into W
            SUBLW .3                  ;TIME – W
            BTFSS    STATUS,ZEROBIT   ;Check TIME-W=0
            GOTO     LOOPB
            RETLW    0                ;Return after TMR0 = 2

;7.5ms SECOND DELAY
DELAY3      CLRF     TMR0             ;Start TMR0
LOOPC       MOVF     TMR0,W           ;Read TMR0 into W
```

```
            SUBLW      .6                    ;TIME – W
            BTFSS      STATUS,ZEROBIT        ;CHECK TIME-W=0
            GOTO       LOOPC
            RETLW      0                     ;Return after TMR0 = 3
```

;CONFIGURATION SECTION

```
START       BSF        STATUS,5              ;Turns to Bank1.
            MOVLW      B'00011111'           ;PORTA is input
            TRIS       PORTA
            MOVLW      B'00000000'
            TRIS       PORTB                 ;PORTB is Output.
            MOVLW      B'00000010'           ;TIMER, 1ms
            OPTION
            BCF        STATUS,5              ;Return to Bank0.
            CLRF       PORTA                 ;Clears PortA.
            CLRF       PORTB                 ;Clears PortB.
```

;***
;Program starts now.

```
BEGIN       BTFSC      PORTA,0               ;wait for switch press
            GOTO       BEGIN
            MOVLW      27H                   ;Put 27H into W
            MOVWF      NUMA                  ;PUT 27H into NUMA

            BCF        PORTB,0
            CALL       DELAY1
            BSF        PORTB,0               ;Transmit START
            CALL       DELAY3                ;wait 7.5ms

TESTA0      BCF        PORTB,0               ;Transmit space
            CALL       DELAY1                ;wait 2.5ms
            BTFSC      NUMA,0                ;Test NUMA,0
            GOTO       SETA0                 ;If NUMA0 = 1
            GOTO       CLRA0                 ;If NUMA0 = 0

SETA0       BSF        PORTB,0               ;Transmit 1
            CALL       DELAY2                ;wait 5ms
            GOTO       TESTA1

CLRA0       BSF        PORTB,0               ;Transmit 0
```

```
                CALL     DELAY1            ;wait 2.5ms

TESTA1          BCF      PORTB,0
                CALL     DELAY1
                BTFSC    NUMA,1
                GOTO     SETA1
                GOTO     CLRA1

SETA1           BSF      PORTB,0
                CALL     DELAY2
                GOTO     TESTA2
CLRA1           BSF      PORTB,0
                CALL     DELAY1

TESTA2          BCF      PORTB,0
                CALL     DELAY1
                BTFSC    NUMA,2
                GOTO     SETA2
                GOTO     CLRA2

SETA2           BSF      PORTB,0
                CALL     DELAY2
                GOTO     TESTA3

CLRA2           BSF      PORTB,0
                CALL     DELAY1

TESTA3          BCF      PORTB,0
                CALL     DELAY1
                BTFSC    NUMA,3
                GOTO     SETA3
                GOTO     CLRA3

SETA3           BSF      PORTB,0
                CALL     DELAY2
                GOTO     TESTA4

CLRA3           BSF      PORTB,0
                CALL     DELAY1

TESTA4          BCF      PORTB,0
                CALL     DELAY1
                BTFSC    NUMA,4
                GOTO     SETA4
```

	GOTO	CLRA4
SETA4	BSF	PORTB,0
	CALL	DELAY2
	GOTO	TESTA5
CLRA4	BSF	PORTB,0
	CALL	DELAY1
TESTA5	BCF	PORTB,0
	CALL	DELAY1
	BTFSC	NUMA,5
	GOTO	SETA5
	GOTO	CLRA5
SETA5	BSF	PORTB,0
	CALL	DELAY2
	GOTO	TESTA6
CLRA5	BSF	PORTB,0
	CALL	DELAY1
TESTA6	BCF	PORTB,0
	CALL	DELAY1
	BTFSC	NUMA,6
	GOTO	SETA6
	GOTO	CLRA6
SETA6	BSF	PORTB,0
	CALL	DELAY2
	GOTO	TESTA7
CLRA6	BSF	PORTB,0
	CALL	DELAY1
TESTA7	BCF	PORTB,0
	CALL	DELAY1
	BTFSC	NUMA,7
	GOTO	SETA7
	GOTO	CLRA7
SETA7	BSF	PORTB,0
	CALL	DELAY2
	CLRF	PORTB

```
                GOTO        BEGIN
CLRA7           BSF         PORTB,0
                CALL        DELAY1
                CLRF        PORTB
                GOTO        BEGIN
END
```

Receiver program code

;RX.ASM

```
TMR0        EQU     1           ;TMR0 is FILE 1.
PORTA       EQU     5           ;PORTA is FILE 5.
PORTB       EQU     6           ;PORTB is FILE 6.
STATUS      EQU     3           ;STATUS is FILE 3.
ZEROBIT     EQU     2           ;ZEROBIT is Bit 2.
CARRY       EQU     0
COUNT       EQU     0CH         ;USER RAM LOCATION.
NUMA        EQU     0DH
;*********************************************************
LIST        P=16F84             ;We are using the 16F84.
ORG         0                   ;0 is the start address.
GOTO        START               ;goto start!

;*********************************************************
;CONFIGURATION SECTION.

START           BSF         STATUS,5        ;Turn to BANK1
                MOVLW       B'00011111'     ;5 bits of PORTA are Inputs.
                TRIS        PORTA
                MOVLW       0
                TRIS        PORTB           ;PORTB is output
                MOVLW       B'00000010'
                OPTION                      ;PRESCALER is /8, 1ms
                BCF         STATUS,5        ;Return to BANK0
                CLRF        PORTA           ;Clears PORTA
                CLRF        PORTB           ;Clears PORTB

;*********************************************************
;Program starts now.

BEGIN           CLRF        NUMA

WAITHI          BTFSS       PORTA,0         ;Wait for HI Transmission
```

```
                GOTO      WAITHI
                CLRF      TMR0
TESTST          BTFSC     PORTA,0         ;Wait for LOW Transmission
                GOTO      TESTST          ;Test for START PULSE
                MOVF      TMR0,W
                SUBLW     .5              ;5-W or 5-TMR0
                BTFSC     STATUS,CARRY    ;SKIP IF TIME>5
                GOTO      WAITHI          ;NOT START BIT
TESTA0H         BTFSS     PORTA,0         ;wait for Hi transmission
                GOTO      TESTA0H
                CLRF      TMR0            ;start timing
TESTA0L         BTFSC     PORTA,0         ;wait for Lo transmission
                GOTO      TESTA0L
                NOP
                MOVF      TMR0,W          ;read value of TMR0
                SUBLW     .3              ;3-W or 3-TMR0
                BTFSS     STATUS,CARRY    ;Is TMR0 > 3 i.e. a logic1
                BSF       NUMA,0          ;Yes, 1 was transmitted.


TESTA1H         BTFSS     PORTA,0         ;Wait for pulse
                GOTO      TESTA1H
                CLRF      TMR0
TESTA1L         BTFSC     PORTA,0         ;Wait for LO.
                GOTO      TESTA1L
                NOP
                MOVF      TMR0,W
                SUBLW     .3
                BTFSS     STATUS,CARRY
                BSF       NUMA,1          ;1 Was transmitted


TESTA2H         BTFSS     PORTA,0         ;Wait for pulse
                GOTO      TESTA2H
                CLRF      TMR0
TESTA2L         BTFSC     PORTA,0         ;Wait for Lo.
                GOTO      TESTA2L
                NOP
                MOVF      TMR0,W
                SUBLW     .3
                BTFSS     STATUS,CARRY
                BSF       NUMA,2          ;1 was transmitted


TESTA3H         BTFSS     PORTA,0         ;Wait for pulse
                GOTO      TESTA3H
```

```
            CLRF      TMR0
TESTA3L     BTFSC     PORTA,0                ;Wait for Lo
            GOTO      TESTA3L
            NOP
            MOVF      TMR0,W
            SUBLW     .3
            BTFSS     STATUS,CARRY
            BSF       NUMA,3                 ;1 was transmitted

TESTA4H     BTFSS     PORTA,0                ;Wait for pulse
            GOTO      TESTA4H
            CLRF      TMR0
TESTA4L     BTFSC     PORTA,0                ;Wait for Lo
            GOTO      TESTA4L
            NOP
            MOVF      TMR0,W
            SUBLW     .3
            BTFSS     STATUS,CARRY
            BSF       NUMA,4                 ;1 was transmitted

TESTA5H     BTFSS     PORTA,0                ;Wait for pulse
            GOTO      TESTA5H
            CLRF      TMR0
TESTA5L     BTFSC     PORTA,0                ;Wait for Lo
            GOTO      TESTA5L
            NOP
            MOVF      TMR0,W
            SUBLW     .3
            BTFSS     STATUS,CARRY
            BSF       NUMA,5                 ;1 was transmitted

TESTA6H     BTFSS     PORTA,0                ;Wait for pulse
            GOTO      TESTA6H
            CLRF      TMR0
TESTA6L     BTFSC     PORTA,0                ;Wait for Lo
            GOTO      TESTA6L
            NOP
            MOVF      TMR0,W
            SUBLW     .3
            BTFSS     STATUS,CARRY
            BSF       NUMA,6                 ;1 was transmitted

TESTA7H     BTFSS     PORTA,0                ;Wait for pulse
            GOTO      TESTA7H
```

```
            CLRF      TMR0
TESTA7L     BTFSC     PORTA,0           ;Wait for Lo
            GOTO      TESTA7L
            NOP
            MOVF      TMR0,W
            SUBLW     .3
            BTFSS     STATUS,CARRY
            BSF       NUMA,7            ;1 was transmitted

            MOVLW     27H
            SUBWF     NUMA,W            ;NUMA-27
            BTFSS     STATUS,ZEROBIT
            GOTO      BEGIN             ;If NUMA is not 27
            BSF       PORTB,0           ;Turn on LED.
            GOTO      BEGIN
END
```

Using the transmit and receive subroutines

The transmit and receive subroutines may seem a little complex, but all you need to do in your code is call them.

- To transmit. Put the data you wish to transmit in the file NUMA then CALL TRANSMIT. The data in the file NUMA is transmitted.
- To receive. CALL RECEIVE, the received data will be present in the file NUMA for you to use.

These programs have illustrated how to switch an LED on (this could be a remote control for a car burglar alarm). You may of course want to add more lines of code to be able to turn the LED off. This could be done in the receiver section by waiting for, say, 2 seconds and on the next transmission turn the LED off, providing of course the code was again 27H. Other codes could of course be added for other switches or keypad buttons, the possibilities are endless.

The transmitter and receiver micros could be hard wired together first to test the software without the radio link. The radio transmitter and receiver can then replace the wire to give a wireless transmission.

10
EEPROM data memory

One of the special features of the 16F84 is the EEPROM data memory. This is a section of memory not in the usual program memory space. It is a block of data like the user files, but unlike the user files the data in the EEPROM data memory is saved when the microcontroller is switched off, i.e. it is non-volatile. Suppose we were counting cars in and out of a car park and we lost the power to our circuit. If we stored the count in EEPROM then we could load our count file with this data and continue without loss of data, when the power returns.

To access the data, i.e. read and write to the EEPROM memory locations, we must of course instruct the microcontroller. There are 64 bytes of EEPROM memory, so we must tell the micro which address we require and if we are reading or writing to it.

When reading we identify the address from 0 to 3Fh using the address register EEADR. The data is then available in register EEDATA. When writing to the EEPROM data we specify the data in the register EEDATA and the location in the register EEADR.

Two other files are used to enable the process, they are EECON1 and EECON2, two EEPROM control registers.

EEPROM registers

Register EECON1 and EECON2 have addresses 8 and 9 respectively in Bank1. The Register EECON1 is below in Figure 10.1.

bit 7	bit 6	bit 5	bit 4	bit 3	bit 2	bit 1	bit 0
-	-	-	EEIF	WRERR	WREN	WR	RD

Figure 10.1 The EECON1 register

Bit 0   RD is set to a 1 to perform a read. It is cleared by the micro when the read is finished.

Bit 1   WR is set to a 1 to perform a write. It is cleared by the micro when the write is finished.

Bit 2   WREN, WRite ENable – a 1 allows the write cycle, a 0 prohibits it.

Bit 3   WRERR reads a 1 if a write is not completed, reads a 0 if the write is completed successfully.

Bit 4   EEIF interrupt flag for the EEDATA – it is a 1 if the write operation is completed, it reads 0 if it is not completed or not started.

Example using the EEPROM

As usual, I think the best way of understanding how this memory works is to look at a simple example.

Suppose we wish to count events, people going into a building, cars going into a car park, etc. If we lose the power to the circuit the data is still retained. The circuit for this is shown in Figure 10.2, the switch press counting circuit.

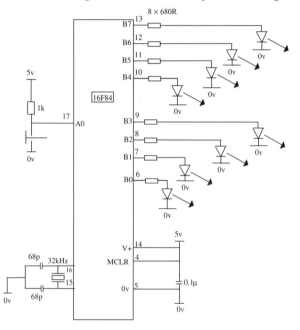

Figure 10.2 Switch press counting circuit

Switch 1 is used to simulate the counting process and the eight LEDs on PORTB display the count in binary. (This is a good chance to practise counting in binary.) The switch of course needs de-bouncing.

Remember the idea of this circuit – we are counting events and displaying the count on PORTB. But if we lose power when the power is re-applied we want to continue the count as if nothing had happened. So when we switch on we must move the previous EEPROM data into the COUNT file.

The flowchart is shown in Figure 10.3.

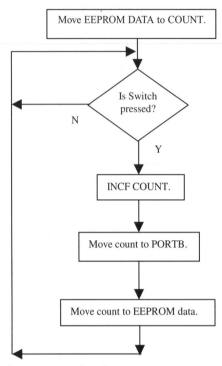

Figure 10.3 The switch press count flowchart

Just a couple of points before we look at the program:

1   We must make sure the EEPROM DATA MEMORY is reset at the very beginning. This is done by writing 00h to EEPROM DATA address 00h when we blow the program into the chip, using the following lines of code.

```
ORG     2100H
DE      00H
```

2100H is the address of the first EEPROM data memory file, i.e. 00h. DE is Define EEPROM data memory, so we are initialising it with 00h, and of course 2101H is EEPROM address1, etc.

2   Reading and writing to EEPROM data is not as straightforward as with user files you probably suspected that! There is a block of code you need to use – just add it to your program as required. When reading EEPROM data at address0 to the file COUNT, CALL READ, the subroutine written in the header. When writing the file COUNT to EEPROM data address0, CALL WRITE.

EEPROM program code

The complete program EEDATAWR.ASM is shown below:

```
;EEDATAWR.ASM    This program will count and display switch
;                presses.
;                The count is saved when the power is removed and con-
;                tinues when the
;                power is re-applied.


TMR0      EQU    1       ;TMR0 is FILE 1.
PORTA     EQU    5       ;PORTA is FILE 5.
PORTB     EQU    6       ;PORTB is FILE 6.
STATUS    EQU    3       ;STATUS is FILE 3.
ZEROBIT   EQU    2       ;ZEROBIT is Bit 2.
COUNT     EQU    0CH     ;USER RAM LOCATION.
EEADR     EQU    9       ;EEPROM address register
EEDATA    EQU    8       ;EEPROM data register
EECON1    EQU    8       ;EEPROM control register1
EECON2    EQU    9       ;EEPROM control register2
RD        EQU    0       ;read bit in EECON1
WR        EQU    1       ;Write bit in EECON1
WREN      EQU    2       ;Write enable bit in EECON1


;*********************************************************
        LIST    P=16C84         ;We are using the 16C84.
        ORG     2100H           ;ADDRESS EEADR 0
        DE      00H             ;put 00H in EEADR 0
        ORG     0               ;0 is the start address.
        GOTO    START           ;goto start!


;*********************************************************
;SUBROUTINE SECTION.

;0.1 SECOND DELAY
DELAYP1   CLRF    TMR0             ;Start TMR0
LOOPA     MOVF    TMR0,W           ;Read TMR0 into W
          SUBLW   .3               ;TIME - W
          BTFSS   STATUS,ZEROBIT   ;CHECK TIME-W=0
          GOTO    LOOPA
          RETLW   0                ;Return after TMR0 = 3

;Put EEDATA 0 into COUNT
```

```
READ       MOVLW    0                  ;read EEDATA from EEADR 0 into W
           MOVWF    EEADR
           BSF      STATUS,5           ;BANK1
           BSF      EECON1,RD
           BCF      STATUS,5           ;BANK0
           MOVF     EEDATA,W
           MOVWF    COUNT
           RETLW    0


;WRITE COUNT INTO EEDATA 0
WRITE      BSF      STATUS,5           ;BANK1
           BSF      EECON1,WREN        ;set WRITE ENABLE
           BCF      STATUS,5           ;BANK0
           MOVF     COUNT,W            ;move COUNT to EEDATA
           MOVWF    EEDATA
           MOVLW    0                  ;set EEADR 0 to receive EEDATA
           MOVWF    EEADR
           BSF      STATUS,5           ;BANK1
           MOVLW    55H                ;55 and AA initiates write cycle
           MOVWF    EECON2
           MOVLW    0AAH
           MOVWF    EECON2
           BSF      EECON1,WR          ;WRITE data to EEADR 0
WRDONE     BTFSC    EECON1,WR
           GOTO     WRDONE             ;wait for write cycle to complete
           BCF      EECON1,WREN
           BCF      STATUS,5           ;BANK0
           RETLW    0


;*********************************************************
;CONFIGURATION SECTION.

START      BSF      STATUS,5           ;Turn to BANK1
           MOVLW    B'00011111'        ;5 bits of PORTA are I/Ps.
           TRIS     PORTA
           MOVLW    0
           TRIS     PORTB              ;PORTB IS OUTPUT
           MOVLW    B'00000111'
           OPTION                      ;PRESCALER is /256
           BCF      STATUS,5           ;Return to BANK0
           CLRF     PORTA              ;Clears PORTA
           CLRF     PORTB              ;Clears PORTB
           CLRF     COUNT
;*********************************************************
```

;Program starts now.

```
              CALL     READ         ;read EEPROM data into COUNT
              MOVF     COUNT,W
              MOVWF    PORTB        ;Display previous COUNT (if any)
PRESS         BTFSC    PORTA,0      ;wait for switch press
              GOTO     PRESS
              CALL     DELAYP1      ;antibounce
RELEASE       BTFSS    PORTA,0      ;wait for switch release
              GOTO     RELEASE
              CALL     DELAYP1      ;antibounce

              INCF     COUNT        ;add 1 to COUNT
              MOVF     COUNT,W      ;put COUNT into W
              MOVWF    PORTB        ;move W (COUNT) to PORTB to display
              CALL     WRITE        ;write COUNT to EEPROM address 0
              GOTO     PRESS        ;return and wait for press
END
```

Microchip are continually expanding their range of microcontrollers and a new series of flash micros has been introduced, namely the 16F87X series which includes 8k of program memory, 368 bytes of user RAM, 256 bytes of EEPROM data memory and an 8 channel 10-bit A/D converter. So now analogue measurements can be stored and saved in EEPROM data!

11
Interrupts

New instructions used in this chapter:

• RETFIE

We all know what interrupts are and we don't like being interrupted. We are busy doing something and the phone rings or someone arrives at the door.

If we are expecting someone, we could look out of the window every now and again to see if they had arrived or we could carry on with what we are doing until the doorbell rings. These are two ways of receiving an interrupt. The first, when we keep checking in software terms, is called polling, the second, when the bell rings, is equivalent to the hardware interrupt.

We have looked at polling when we used the keypad to see if any keys had been pressed. We will now look at the interrupt generated by the hardware.

Before moving onto an example of an interrupt consider the action of the door in a washing machine. The washing cycle does not start until the door is closed, but after that the door does not take any part in the program. But what if a child opens the door, water could spill out or worse! We need to switch off the outputs if the door is opened. To keep looking at the door at frequent intervals in the program (software polling) would be very tedious indeed, so we use a hardware interrupt. We carry on with the program and ignore the door. But if the door is opened the interrupt switches off the outputs – spin motor, etc. If the door had been opened accidentally then closing the door would return to the program for the cycle to continue.

This suggests that when an interrupt occurs we need to remember what the contents of the files were, i.e. the status register, W register, TMR0 and PORT settings, so that when we return from the interrupt the settings are restored. If we did not remember the settings, we could not continue where we left off, because the interrupt switches off all the outputs and the W register would also be altered, at the very least.

Interrupt sources

The 16F84 has four interrupt sources:

- Change on rising or falling edge of PORTB,0.
- TMR0 overflowing from FFh to 00h.
- PORTB bits 4–7 changing.
- DATA EEPROM write complete.

These interrupts can be enabled or disabled as required by their own interrupt enable/disable bits. These bits can be found in the interrupt control register, INTCON.

Interrupt control register

The interrupt control register, INTCON, file 0Bh, is shown in Figure 11.1.

bit 7	bit 6	bit 5	bit 4	bit 3	bit 2	bit 1	bit 0
GIE	EEIE	T0IE	INTE	RBIE	T0IF	INTF	RBIF

Figure 11.1 The interrupt control register, INTCON

Before any of the individual enable bits can be switched ON, the Global Interrupt Enable (GIE) bit 7 must be set, i.e. a 1 enables all unmasked interrupts and a 0 disables all interrupts.

Bit 6   EEIE is an EEPROM data write complete interrupt enable bit, a 1 enables this interrupt and a 0 disables it.

Bit 5   T0IE is the TMR0 overflow interrupt enable bit, a 1 enables this interrupt and a 0 disables it.

Bit 4   INTE is the RB0/INT interrupt enable bit, a 1 enables this interrupt and a 0 disables it.

Bit 3   RBIE is the RB PORT change (B4–B7) interrupt enable bit, a 1 enables it and a 0 disables it.

Bit 2   T0IF is the flag, which indicates TMR0 has overflowed to generate the interrupt. 1 indicates TMR0 has overflowed, 0 indicates it hasn't. This bit must be cleared in software.

Bit 1   INTF is the RB0/INT interrupt flag bit which indicates a change on PORTB,0. A 1 indicates a change has occurred, a 0 indicates it hasn't.

Bit 0   RBIF is the RB PORT Change interrupt flag bit. A 1 indicates that one of the inputs PORTB,4–7 has changed state. This bit must be cleared in software. A 0 indicates that none of the PORTB,4-7 bits have changed.

Program using an interrupt

As an example of how an interrupt works consider the following example. Suppose we have four lights flashing consecutively for 5 seconds each. A switch connected to B0 acts as an interrupt so that when B0 is at a logic0 an interrupt routine is called. This interrupt routine flashes all 4 lights ON and OFF twice at

1 second intervals and then returns to the program providing the switch on B0 is at a logic1.

The circuit diagram for this application is shown in Figure 11.2.

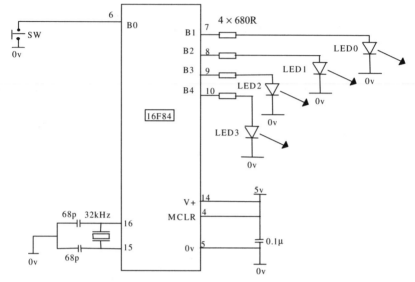

Figure 11.2 Interrupt demonstration circuit

One thing to note from the circuit is that the 16F84 chip has internal pull-up resistors on PORTB so B0 does not need a pull-up resistor on the switch.

The interrupt we are using is a change on B0, we are therefore concerned with the following bits in the INTCON register, i.e. INTE bit4 the enable bit and INTF bit1 the flag showing B0 has changed, and of course GIE bit7 the Global Interrupt Enable bit.

Program operation

When B0 generates an interrupt the program branches to the interrupt service routine. Where? Program memory location 4 tells the microcontroller where to go to find the interrupt service routine. Program memory location 4 is then programmed using the org statement as:

```
ORG    4      ;write next instruction in program memory location 4
GOTO   ISR    ;jump to the Interrupt Service Routine.
```

The interrupt service routine

The interrupt service routine, ISR, is written like a subroutine and is shown below:

;Interrupt Service Routine.

```
          MOVWF     W_TEMP        ;Save W
          MOVF      STATUS,W
          MOVWF     STATUS_T      ;Save STATUS
          MOVF      TMR0,W
          MOVWF     TMR0_T        ;Save TMR0
          MOVF      PORTB,W
          MOVWF     PORTB_T       ;Save PORTB

          MOVLW     0FFH
          MOVWF     PORTB         ;turn on all outputs.
          CALL      DELAPY1       ;1 second delay
          MOVLW     0
          MOVWF     PORTB         ;turn off all outputs
          CALL      DELAPY1       ;1 second delay
          MOVLW     0FFH
          MOVWF     PORTB         ;turn on all outputs.
          CALL      DELAPY1       ;1 second delay
          MOVLW     0
          MOVWF     PORTB         ;turn off all outputs
          CALL      DELAPY1       ;1 second delay
SW_HI     BTFSS     PORTB,0
          GOTO      SW_HI         ;wait for switch to be HI.

          MOVF      STATUS_T,W
          MOVWF     STATUS        ;Restore STATUS
          MOVF      TMR0_T,W
          MOVWF     TMR0          ;Restore TMR0
          MOVF      PORTB_T,W
          MOVWF     PORTB         ;Restore PORTB
          MOVF      W_TEMP,W      ;Restore W

          BCF       INTCON,INTF   ;Reset Interrupt Flag
          RETFIE                  ;Return from the interrupt
```

Operation of the interrupt service routine

The interrupt service routine operates in the following way.

- When an interrupt is made the Global Interrupt Enable is cleared automatically (disabled) to switch off all further interrupts. We would not wish to be interrupted while we are being interrupted.
- The registers W, STATUS, TMR0 and PORTB are saved in temporary locations W_TEMP, STATUS_T, TMR0_T and PORTB_T.
- The interrupt routine is executed, the lights flash on and off twice. This is a separate sequence than before to show the interrupt has interrupted the normal flow of the program. Note: The program has not been looking at the switch that generated the interrupt.
- We then wait until the switch returns HI.
- The temporary files W_TEMP, STATUS_T, TMR0_T and PORTB_T are restored back into W, STATUS, TMR0 and PORTB.
- The PORTB,0 interrupt flag INTCON,INTF is cleared ready to indicate further interrupts.
- We return from the interrupt, and the Global Interrupt Enable bit is automatically set to enable further interrupts.

Program of the interrupt demonstration

The complete code for this program is shown below as INTFLASH.ASM.

;INTFLASH.ASM Flashing lights being interrupted by a switch on B0.

;EQUATES SECTION

TMR0	EQU	1	;means TMR0 is file 1.
STATUS	EQU	3	;means STATUS is file 3.
PORTA	EQU	5	;means PORTA is file 5.
PORTB	EQU	6	;means PORTB is file 6.
INTCON	EQU	0BH	;Interrupt Control Register
ZEROBIT	EQU	2	;means ZEROBIT is bit 2.
GIE	EQU	7	;Global Interrupt bit
INTE	EQU	4	;B0 interrupt enable bit.
INTF	EQU	1	;B0 interrupt flag
COUNT	EQU	0CH	;means COUNT is file 0C, ;a register to count events.
TMR0_T	EQU	0DH	;TMR0 temporary file
W_TEMP	EQU	0EH	;W temporary file
STATUS_T	EQU	0FH	;STATUS temporary file
PORTB_T	EQU	10H	;PORTB temporary file

;***

LIST	P=16F84	;we are using the 16F84.
ORG	0	;the start address in memory is 0

```
GOTO      START            ; goto start!
ORG       4
GOTO      ISR              ;location4 jumps to ISR
;*********************************************************
;

;SUBROUTINE SECTION.

;1 second delay.
DELAY1    CLRF    TMR0              ;START TMR0.
LOOPA     MOVF    TMR0,W            ;READ TMR0 INTO W.
          SUBLW   .32               ;TIME - 32
          BTFSS   STATUS,ZEROBIT    ;Check TIME-W = 0
          GOTO    LOOPA             ;Time is not = 32.
          RETLW   0                 ;Time is 32, return.

;5 second delay.
DELAY5    CLRF    TMR0              ;START TMR0.
LOOPB     MOVF    TMR0,W            ;READ TMR0 INTO W.
          SUBLW   .160              ;TIME - 160
          BTFSS   STATUS,ZEROBIT    ;Check TIME-W = 0
          GOTO    LOOPB             ;Time is not = 160.
          RETLW   0                 ;Time is 160, return.

;Interrupt Service Routine.

ISR       MOVWF   W_TEMP            ;Save W
          MOVF    STATUS,W
          MOVWF   STATUS_T          ;Save STATUS
          MOVF    TMR0,W
          MOVWF   TMR0_T            ;Save TMR0
          MOVF    PORTB,W
          MOVWF   PORTB_T           ;Save PORTB

          MOVLW   0FFH
          MOVWF   PORTB             ;turn on all outputs.
          CALL    DELAY1            ;1 second delay
          MOVLW   0
          MOVWF   PORTB             ;turn off all outputs
          CALL    DELAY1            ;1 second delay
          MOVLW   0FFH
          MOVWF   PORTB             ;turn on all outputs.
          CALL    DELAY1            ;1 second delay
          MOVLW   0
          MOVWF   PORTB             ;turn off all outputs
```

```
            CALL      DELAY1              ;1 second delay
SW_HI       BTFSS     PORTB,0
            GOTO      SW_HI               ;wait for switch to be HI.

            MOVF      STATUS_T,W
            MOVWF     STATUS              ;Restore STATUS
            MOVF      TMR0_T,W
            MOVWF     TMR0                ;Restore TMR0
            MOVF      PORTB_T,W
            MOVWF     PORTB               ;Restore PORTB
            MOVF      W_TEMP,W            ;Restore W

            BCF       INTCON,INTF         ;Reset Interrupt Flag
            RETFIE                        ;Return from the interrupt
```

;**
;CONFIGURATION SECTION

```
START       BSF       STATUS,5            ;Turns to Bank1.
            MOVLW     B'00011111'         ;5bits of PORTA are I/P
            TRIS      PORTA
            MOVLW     B'00000001'
            TRIS      PORTB               ;B0 is an input, B1-7 O/P
            MOVLW     B'00000111'         ;Prescaler is /256
            OPTION                        ;TIMER is 1/32 secs.
            BCF       STATUS,5            ;Return to Bank0.
            CLRF      PORTA               ;Clears PortA.
            CLRF      PORTB               ;Clears PortB.
            BSF       INTCON,GIE          ;Enable Global Interrupt
            BSF       INTCON,INTE         ;Enable B0 interrupt
```

;**
;Program starts now.

```
BEGIN       MOVLW     B'00000010'         ;Turn on B1
            MOVWF     PORTB
            CALL      DELAY5              ;wait 5 seconds
            MOVLW     B'00000100'         ;Turn on B2
            MOVWF     PORTB
            CALL      DELAY5              ;wait 5 seconds
            MOVLW     B'00001000'         ;Turn on B3
            MOVWF     PORTB
            CALL      DELAY5              ;wait 5 seconds
            MOVLW     B'00010000'         ;Turn on B4
```

```
        MOVWF    PORTB
        CALL     DELAY5        ;wait 5 seconds
        GOTO     BEGIN
END
```

The four lights are flashing on and off slowly enough (5 second intervals) so that you can interrupt part way through taking B0 low via the switch (make sure B0 is hi when starting). The interrupt service routine then flashes all the lights on and off twice at 1 second intervals.

When returning from the interrupt with B0 hi again, the program resumes from where it left off, i.e. if the second LED had been on for 3 seconds it would come back on for the remaining 2 seconds and the sequence would continue.

12
The 12C5XX series 8 pin microcontroller

Arizona Microchip has introduced a range of microcontrollers with 8 pins. They include types with data EEPROM and A/D converters. In this section we will cover the basic devices, the 12C508 and 12C509.

The differences in the two types are shown in Table 12.1.

Table 12.1 12C508/509 Memory

Device	EEPROM	User Files	Registers
12C508	512×12	25	7
12C509	1024×12	41	7

Pin diagram of the 12C508/509

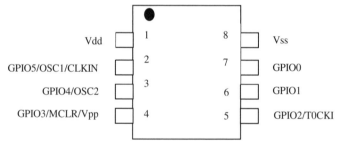

Figure 12.1 12C58/509 pin diagram

Features of the 12C50X

One of the special features of this micro is that it has eight pins, but six of them can be used as I/O pins, the remaining two pins being used for the power supply. There is no need to add a crystal and capacitors, because a 4MHz oscillator is built on board! If you wish to use a clock other than the 4MHz provided, then you can connect an oscillator circuit to pins 2 and 3 (as in the 16F84). That leaves you with of course only four I/O.

Being an 8 pin device means of course it is smaller than an 18 pin device and cheaper! The onboard oscillator means that the crystal and timing capacitors are no longer required, reducing the component count, size and cost even further. So if your application requires no more than six I/O this device is the one for you. It has useful applications in burglar alarm circuits and the radio transmitter circuits we have looked at previously.

The memory map of the 12C508

The memory map of the 12C508 is shown in Figure 12.2, showing the seven registers and 25 user files.

Address	File
01h	TMR0
02h	PCL
03h	STATUS
04h	FSR
05h	OSCCAL
06h	GPIO
07h	
	General Purpose Registers (user files)
1Fh	

Figure 12.2 12C508 memory map

The 12C509 has 16 extra user files mapped in Bank1.

There is no longer a PORTA or PORTB because we only have six I/O, they are in a port called GPIO (General Purpose Input Output), File 6.

Oscillator calibration

Apart from the small size of this device an appealing feature is that the oscillator is on board. A new addition is file 5, OSCCAL. This is an oscillator calibration file used to trim the 4MHz oscillator.

The 4MHz oscillator takes its timing from an onboard R-C network, which is not very precise. So every chip has a value that can be put into OSCCAL to trim it. This value is stored in the last memory address, i.e. 01FFh for the 12C508 and 03FFh for the 12C509. The code, which is loaded by the manufacturer in the last memory location, is MOVLW XX where XX is the trimming value. The last memory location is the reset vector, i.e. when switched on the micro goes to this location first, it loads the calibration value into W and the program counter overflows to 000h and continues executing the code. To use the calibration value, in the Configuration Section write the instruction MOVWF OSC-CAL, which then moves the manufacturer's calibration value into the timing circuit. There is one point to remember – if you are using a windowed device then the calibration value will be erased when the memory is erased. So make a note of the MOVLW XX code and program it back in. The trimming can be ignored if required – but it only requires one line of code, so why not use it.

I/O PORT, GPIO

The GPIO, General Purpose Input/Output, is an 8-bit I/O register, it has six I/O lines available so bits GPIO 0 to 5 are used, bits 6 and 7 are not. Note: GPIO bit 3 is an input only pin so there is a maximum of five outputs. GPIO pins 0, 1 and 3 can be configured with weak pull-ups by writing 0 to OPTION,6 (bit 6 in the OPTION register).

Delays with the 12C508/509

We have previously used a 32kHz crystal with the 16F84 device, but now we are going to use the internal 4MHz clock with the 12C508. A 4MHz clock means that the basic timing is ¼ of this, i.e. 1MHz. If we set the OPTION register to divide by 256 this gives a timing frequency of 3906Hz. In the header for the 12C508 I have (as with the 16F84) included a 1 second and a 0.5 second delay. In order to achieve a 1 second delay from a frequency of 3906Hz I first produced a delay of 1/100 second by counting 39 timing pulses, i.e. 3906Hz/39 = 100.15 = 100Hz approx., called DELAY. A 1 second delay, subroutine DELAY1, then counts 100 of these DELAY times (i.e. 100 × 1/100 second), and of course a delay of 0.5 seconds would count 50. Just before we look at the header – we do not have an instruction SUBLW on the 12C508. I have therefore set up a file called TIME that I have written 39 into. I then move TMR0 into W and subtract the file TIME (39d) from it to see if TMR0 = 39, i.e. 1/100 of a second has elapsed.

WARNING: The 12C508 and 509 micros only have a two level deep stack. Which means when you do, for example, a 1 second delay, CALL DELAY1, this then calls another subroutine, i.e. CALL DELAY. You have used your two levels and cannot do any further calls without returning from at least one of those subroutines. If you did make a third CALL the program would not be able to find its way back!

Header for 12C508/509

```
;HEADER12.ASM FOR 12C508/9.

TMR0      EQU    1          ;TMR0 is FILE 1.
OSCCAL    EQU    5          ;Oscillator calibration
GPIO      EQU    6          ;GPIO is FILE 6.
STATUS    EQU    3          ;STATUS is FILE 3.
ZEROBIT   EQU    2          ;ZEROBIT is Bit 2.
COUNT     EQU    07H        ;USER RAM LOCATION.
TIME      EQU    08H        ;TIME IS 39
;************************************************************
LIST P=12C508              ;We are using the 12C508.
ORG 0                      ;0 is the start address.
GOTO START                 ;goto start!

;************************************************************
;SUBROUTINE SECTION.


DELAY     CLRF    TMR0              ;Start TMR0
LOOPA     MOVF    TMR0,W            ;Read TMR0 into W
          SUBWF   TIME,W            ;TIME – W
          BTFSS   STATUS,ZEROBIT    ;Check TIME-W=0
          GOTO    LOOPA
          RETLW   0                 ;Return after TMR0 = 39

;1 SECOND DELAY
DELAY1    MOVLW   .100
          MOVWF   COUNT
TIMEA     CALL    DELAY
          DECFSZ  COUNT
          GOTO    TIMEA
          RETLW   0

;1/2 SECOND DELAY
DELAYP5   MOVLW   .50
          MOVWF   COUNT
TIMEB     CALL    DELAY
          DECFSZ  COUNT
          GOTO    TIMEB
          RETLW   0

;************************************************************
```

```
;CONFIGURATION SECTION.

START      MOVWF    OSCCAL           ;Calibrate oscillator.
           MOVLW    B'00001000'      ;5 bits of GPIO are O/Ps.
           TRIS     GPIO             ;Bit 3 is Input
           MOVLW    B'00000111'
           OPTION                    ;PRESCALER is /256
           CLRF     GPIO             ;Clear GPIO
           MOVLW    .39
           MOVWF    TIME             ;TIME=39
;**********************************************************
;Program starts now.
```

Program application for 12C508

There are five outputs on the 12C508, i.e. GPIO bits 0, 1, 2, 4 and 5. Bit 3 is an input only. For our application we will chase five LEDs on our outputs backwards and forwards at 0.5 second intervals. The circuit diagram is shown in Figure 12.3.

Notice that the only other component required is the power supply decoupling capacitor, 0.1µF, no oscillator circuit is required.

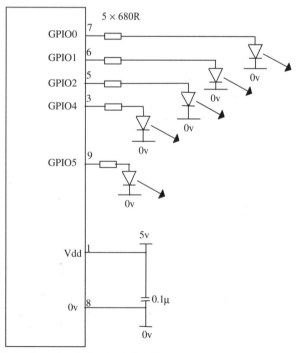

Figure 12.3 LED chasing circuit for the 12C508

The program for the LED chasing project, LED_CH12.ASM

;LED_CH12.ASM Program to chase 5 LEDs with the 12C508

```
TMR0          EQU          1          ;TMR0 is FILE 1.
GPIO          EQU          6          ;GPIO is FILE 6.
STATUS        EQU          3          ;STATUS is FILE 3.
ZEROBIT       EQU          2          ;ZEROBIT is Bit 2.
COUNT         EQU          07H        ;USER RAM LOCATION.
TIME          EQU          08H        ;TIME IS 39
;********************************************************
LIST          P=12C508                ;We are using the 12C508.
ORG           0                       ;0 is the start address.
GOTO          START                   ;goto start!


;********************************************************
;SUBROUTINE SECTION.


DELAY         CLRF         TMR0                  ;Start TMR0
LOOPA         MOVF         TMR0,W                ;Read TMR0 into W
              SUBWF        TIME,W                ;TIME – W
              BTFSS        STATUS,ZEROBIT        ;Check TIME-W=0
              GOTO         LOOPA
              RETLW        0                     ;Return after TMR0 = 39

;1 SECOND DELAY
DELAY1        MOVLW        .100
              MOVWF        COUNT
TIMEA         CALL         DELAY
              DECFSZ       COUNT
              GOTO         TIMEA
              RETLW        0

;1/2 SECOND DELAY
DELAYP5       MOVLW        .50
              MOVWF        COUNT
TIMEB         CALL         DELAY
              DECFSZ       COUNT
              GOTO         TIMEB
              RETLW        0


;********************************************************
```

;CONFIGURATION SECTION.

```
START          MOVWF     OSCCAL              ;Calibrate oscillator.

               MOVLW     B'00001000'         ;5 bits of GPIO are O/Ps.
               TRIS      GPIO                ;Bit 3 is Input
               MOVLW     B'00000111'
               OPTION                        ;PRESCALER is /256
               CLRF      GPIO                ;Clear GPIO
               MOVLW     .39
               MOVWF     TIME                ;TIME=39
;*********************************************************
;Program starts now.
BEGIN          MOVLW     B'00000001'         ;turn on LED0
               MOVWF     GPIO
               CALL      DELAYP5
               MOVLW     B'00000010'         ;turn on LED1
               MOVWF     GPIO
               CALL      DELAYP5
               MOVLW     B'00000100'         ;turn on LED2
               MOVWF     GPIO
               CALL      DELAYP5
               MOVLW     B'00010000'         ;turn on LED3
               MOVWF     GPIO
               CALL      DELAYP5
               MOVLW     B'00100000'         ;turn on LED4
               MOVWF     GPIO
               CALL      DELAYP5
               MOVLW     B'00010000'         ;turn on LED3
               MOVWF     GPIO
               CALL      DELAYP5
               MOVLW     B'00000100'         ;turn on LED2
               MOVWF     GPIO
               CALL      DELAYP5
               MOVLW     B'00000010'         ;turn on LED1
               MOVWF     GPIO
               CALL      DELAYP5
               GOTO      BEGIN
END
```

The program is similar in content to the 16F84 programs used previously, but with the following exceptions:

- A file TIME, file 8, has been set up which has had 39 loaded into it, in the Configuration Section. This is used to determine when TMR0 has reached a count of 39, a time of 0.01 seconds, which is then used in the timing subroutines.
- In the Configuration Section the first instruction the program encounters is MOVWF OSCCAL. This moves the calibration value which has just been read by MOVLW XX, from location 1FFH, the first instruction, into the calibration file OSCCAL.
- GPIO is used in the program instead of the usual PORTA and PORTB.

13
The 16F87X microcontroller

PIC manufacturers, Arizona Microchip, are continually expanding their range of microcontrollers. They have recently introduced a range of flash devices (no eraser required), with the family name 16F87X. This range includes the devices 16F870, 16F871, 16F872, 16F873, 16F874, 16F876 and 16F877. They are basically the same device but differ in the amounts of I/O, analogue inputs, program memory, data memory (RAM) and EEPROM data memory that they have.

The 16F87X range is similar to the 16F84 and 16C711 combined – but bigger and better. They have more I/O, program memory, data memory, and EEPROM data memory than the 16F84, and have more analogue inputs than the 16C711 and use 10-bit A/D conversion instead of 8 bits. So the A/D resolution is 1024 instead of 256.

16F87X family specification

Device	Program Memory	EEPROM Data Memory (bytes)	RAM (bytes)	Pins	I/O	10-bit A/D Channels
16F870	2k	64	128	28	22	5
16F871	2k	64	128	40	33	8
16F872	2k	64	128	28	22	5
16F873	4k	128	192	28	22	5
16F874	4k	128	192	40	33	8
16F876	8k	256	368	28	22	5
16F877	8k	256	368	40	33	8

16F87X memory map

The 16F87X devices have more functions than we have seen previously. These functions of course need registers in order to make the various selections.

The memory map of the 16F87X showing these registers is shown in Figure 13.1.

The 16F87X devices have a number of extra registers that are not required in the applications we have looked at. For an explanation of these registers please

Address	File Name	File Name	File Name	File Name
00h	Ind.Add	Ind.Add	Ind.Add	Ind.Add
01h	TMR0	Option	TMR0	Option
02h	PCL	PCL	PCL	PCL
03h	Status	Status	Status	Status
04h	FSR	FSR	FSR	FSR
05h	PORTA	TRISA		
06h	PORTB	TRISB	PORTB	TRISB
07h	PORTC	TRISC		
08h	PORTD	TRISD		
09h	PORTE	TRISE		
0Ah	PCLATH	PCLATH	PCLATH	PCLATH
0Bh	INTCON	INTCON	INTCON	INTCON
0Ch	PIR1	PIE1	EEDATA	EECON1
0Dh	PIR2	PIE2	EEADR	EECON2
0Eh	TMR1L	PCON	EEDATH	
0Fh	TMR1H		EEADRH	
10h	T1CON			
11h	TMR2	SSPCON2		
12h	T2CON	PR2		
13h	SSPBUF	SSPADD		
14h	SSPCON	SSPSTAT		
15h	CCPR1L			
16h	CCPR1H			
17h	CCP1CON		General	General
18h	RCSTA	TXSTA	Purpose	Purpose
19h	TXREG	SPBRG	Register	Register
1Ah	RCREG		96 bytes	96 bytes
1Bh	CCPR2L			
1Ch	CCPR2H			
1Dh	CCP2CON			
1Eh	ADRESH	ADRESL		
1Fh	ADCON0	ADCON1		
·	General	General		
	Purpose	Purpose		
	Register	Register		
6Fh	96 bytes	80 bytes		
·				
7Fh				
	Bank0	Bank1	Bank2	Bank3

Figure 13.1 16F87X memory map

see Microchip's website www.microchip.com where you can download the data sheet as a pdf (portable document file) which can be read using Adobe Acrobat Reader.

The 16F872 microcontroller

In order to demonstrate the operation of the 16F87X series we will consider the 16F872 device. This is a 28 pin device with 22 I/O available on three ports. PortA has six I/O, PortB has eight I/O and PORTC has eight I/O. Of the six I/O available on PortA five of them can be analogue inputs. The header for the 16F872, HEAD872.ASM, configures the device with five analogue inputs on PortA, eight digital inputs on PortC and eight outputs on PortB. The Port Configuration for the device is shown in Figure 13.2.

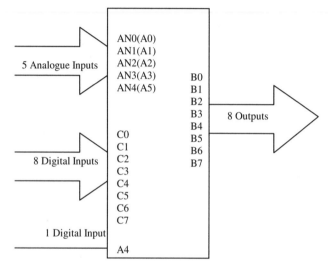

Figure 13.2 Port Configuration of the 16F872

The 16F872 has been configured in HEADER872.ASM to allow all the programs used previously to be copied over with as little alteration as possible.

The 16F872 Header

;HEAD872.ASM

;EQUATES SECTION

TMR0	EQU	1
OPTION_R	EQU	1
PORTA	EQU	5
PORTB	EQU	6
PORTC	EQU	7
TRISA	EQU	5
TRISB	EQU	6

```
                TRISC      EQU        7
                STATUS     EQU        3
                ZEROBIT    EQU        2
                CARRY      EQU        0
                EEADR      EQU        0DH
                EEDATA     EQU        0CH
                EECON1     EQU        0CH
                EECON2     EQU        0DH
                RD         EQU        0
                WR         EQU        1
                WREN       EQU        2
                ADCON0     EQU        1FH
                ADCON1     EQU        1FH
                ADRES      EQU        1EH
                CHS0       EQU        3
                GODONE     EQU        2
                COUNT      EQU        20H
;***************************************************
                LIST       P=16F872
                ORG        0
                GOTO       START
;***************************************************
;SUBROUTINE SECTION.

;1 SECOND DELAY
DELAY1          CLRF       TMR0                     ;Start TMR0
LOOPA           MOVF       TMR0,W                   ;Read TMR0 into W
                SUBLW      .32                      ;TIME – W
                BTFSS      STATUS,ZEROBIT           ;Check TIME-W=0
                GOTO       LOOPA
                RETLW      0                        ;Return after TMR0 = 32

;0.5 SECOND DELAY
DELAYP5         CLRF       TMR0                     ;Start TMR0
LOOPB           MOVF       TMR0,W                   ;Read TMR0 into W
                SUBLW      .16                      ;TIME – W
                BTFSS      STATUS,ZEROBIT           ;Check TIME-W=0
                GOTO       LOOPB
                RETLW      0                        ;Return after TMR0 = 16

;**********************************************************
;CONFIGURATION SECTION.

START           BSF        STATUS,5                 ;Bank1
```

```
MOVLW       B'11111111'
MOVWF       TRISA           ;PortA is input

MOVLW       B'00000000'
MOVWF       TRISB           ;PortB is output

MOVLW       B'11111111'
MOVWF       TRISC           ;PortC is input

MOVLW       B'00000111'
MOVWF       OPTION_R        ;Option Register, TMR0 / 256

MOVLW       B'00000000'
MOVWF       ADCON1          ;PortA bits 0, 1, 2, 3, 5 are analogue
BSF         STATUS,6        ;BANK3
BCF         EECON1,7        ;Data memory on.
BCF         STATUS,5
BCF         STATUS,6        ;BANK0 return
BSF         ADCON0,0        ;turn on A/D.
CLRF        PORTA
CLRF        PORTB
CLRF        PORTC
```

;***
;Program starts now.

Explanation of HEAD872.ASM

Equates Section

- The first difference here is that the OPTION register now has a file address, it is file 1 in Bank1.
- We have a third port, PORTC file 7, and its corresponding TRIS file, TRISC file 7 on Bank1. The TRIS file sets the I/O direction of the port bits.
- The EEPROM data file addresses have been included. EEADR is file 0Dh in Bank2, EEDATA is file 0Ch in Bank2, EECON1 is file 0Ch in Bank3 and EECON2 is file 0Dh in Bank3.
- The EEPROM data bits have been added. RD the read bit is bit 0, WR the write bit is bit 1, WREN the write enable bit is bit 2.
- The analogue files ADRES, ADCON1 and ADCON0 have been included as have the associated bits CHS0 channel 0 select, bit 3, and the GODONE bit, bit 2.

List Section

- This of course indicates the microcontroller being used, the 16F872, and that the first memory location is 0. In address 0 is the instruction GOTO START that instructs the micro to bypass the subroutine section and goto the Configuration Section at the label START.

Subroutine Section

- This includes the two delays DELAY1 and DELAYP5 as before.

Configuration Section

- As before we need to switch to Bank1 to address the TRIS files to configure the I/O. PORTA is set as an input port with the two instructions

```
MOVLW    B'11111111'
MOVWF    TRISA
```

- The recommended command now uses the file TRISA rather than the previous TRIS PORTA. PORTB and PORTC are configured in a similar manner using TRISB and TRISC.
- The OPTION register is configured with the instructions

```
MOVLW    B'00000111'
MOVWF    OPTION_R
```

These instructions now use the file name OPTION_R to move the data into the Option Register instead of the previous command OPTION.

- The A/D register is configured with the instructions

```
MOVLW    B'00000000'
MOVWF    ADCON1
```

Setting PORTA bits 0, 1, 2, 3 and 5 as analogue inputs.

- We turn to Bank3 by setting Bank select bit, STATUS,6 (bit 5 is still set) so that we can address EECON1, the EEPROM data control register. BSF EECON1 then enables access to the EEPROM program memory when required.
- We then turn back to Bank0 by clearing bits 5 and 6 of the status register and clear the files PortA, PortB and PortC.

16F872 application – a greenhouse control

In order to demonstrate the operation of the 16F872 and to develop our programming skills a little further consider the following application.

- A greenhouse has its temperature monitored so that a heater is turned on when the temperature drops below 15°C and turns the heater off when the temperature is above 17°C.
- A probe in the soil monitors the soil moisture so that a water valve will open for 5 seconds to irrigate the soil if it dries out. The valve is closed and will not be active for a further 5 seconds to give the water time to drain into the soil.
- A float switch monitors the level of the water and sounds an alarm if the water drops below a minimum level.

The circuit diagram for the greenhouse control is shown in Figure 13.3 and the flowchart is drawn in Figure 13.4.

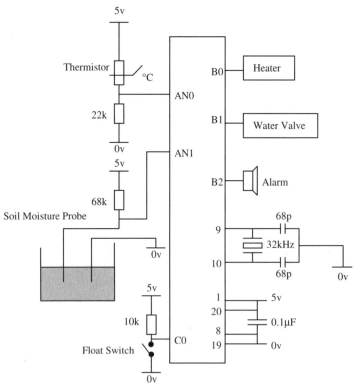

Figure13.3 Greenhouse control circuit

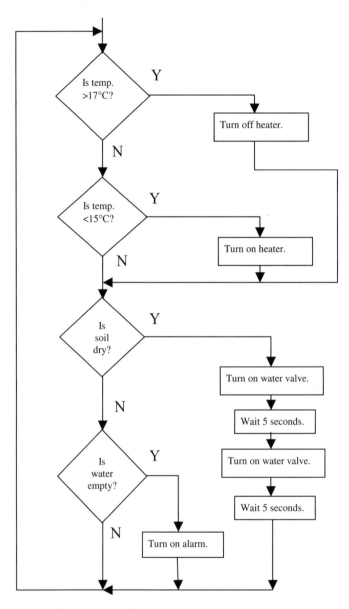

Figure 13.4 Greenhouse control flowchart

Greenhouse program

In order to program the analogue/digital settings consider the NTC thermistor. As the temperature increases the resistance of the thermistor will decrease and so the voltage presented to AN0 will increase. Let us assume the voltage is 2.9v at 15°C and 3.2v at 17°C – they correspond to digital readings of 2.9 × 51 = 147.9, i.e. 148 and 3.2 × 51 = 163.2, i.e. 163. Note: 5v = 255, so 1v = 51 – we are using an 8-bit A/D. Our program then needs to check when AN0 goes above 163 and below 148.

As the soil dries out its resistance will increase. Let us assume in our application dry soil will give a reading of 2.6v (on AN1), i.e. 2.6 × 51 = 132.6, i.e. 133. So any reading above 133 is considered dry.

The float switch is a digital input showing 1 if the water level is above the minimum required and a 0 if it is below the minimum.

Greenhouse code

The code for the greenhouse uses HEAD872.ASM with the program instuctions added and saved as GREENHO.ASM.

```
;GREENHO.ASM
;EQUATES SECTION

            TMR0        EQU     1
            OPTION_R    EQU     1
            PORTA       EQU     5
            PORTB       EQU     6
            PORTC       EQU     7
            TRISA       EQU     5
            TRISB       EQU     6
            TRISC       EQU     7
            STATUS      EQU     3
            ZEROBIT     EQU     2
            CARRY       EQU     0
            EEADR       EQU     0DH
            EEDATA      EQU     0CH
            EECON1      EQU     0CH
            EECON2      EQU     0DH
            RD          EQU     0
            WR          EQU     1
            WREN        EQU     2
            ADCON0      EQU     1FH
            ADCON1      EQU     1FH
            ADRES       EQU     1EH
            CHS0        EQU     3
            GODONE      EQU     2
            COUNT EQU           20H
```

```
;*****************************************************
        LIST        P=16F872
        ORG         0
        GOTO        START
;*****************************************************
;SUBROUTINE SECTION.

;1 SECOND DELAY
DELAY1      CLRF        TMR0            ;Start TMR0
LOOPA       MOVF        TMR0,W          ;Read TMR0 into W
            SUBLW       .32             ;TIME - W
            BTFSS       STATUS,ZEROBIT  ;Check TIME-W=0
            GOTO        LOOPA
            RETLW       0               ;Return after TMR0 = 32

;0.5 SECOND DELAY
DELAYP5     CLRF        TMR0            ;Start TMR0
LOOPB       MOVF        TMR0,W          ;Read TMR0 into W
            SUBLW       .16             ;TIME - W
            BTFSS       STATUS,ZEROBIT  ;Check TIME-W=0
            GOTO        LOOPB
            RETLW       0               ;Return after TMR0 = 16

;5 SECOND DELAY
DELAY5      CLRF        TMR0            ;Start TMR0
LOOPC       MOVF        TMR0,W          ;Read TMR0 into W
            SUBLW       .160            ;TIME - W
            BTFSS       STATUS,ZEROBIT  ;Check TIME-W=0
            GOTO        LOOPC
            RETLW       0               ;Return after TMR0 = 160


HEAT_ON     BSF         PORTB,0         ;Turn heater on
            GOTO        SOIL            ;Check soil moisture

HEAT_OFF    BCF         PORTB,0         ;Turn heater off
            GOTO        SOIL            ;Check soil moisture

WATER_ON    BSF         PORTB,1         ;Turn water on
            CALL        DELAY5
            BCF         PORTB,1         ;Turn water off
            CALL        DELAY5
            GOTO        WATER           ;Check water level
ALARM_ON    BSF         PORTB,2         ;Turn alarm on
```

```
                GOTO        BEGIN              ;Repeat the process
ALARM_OFF   BCF         PORTB,2            ;Turn alarm off
                GOTO        BEGIN              ;Repeat the process
```

;***
;CONFIGURATION SECTION.

```
START       BSF         STATUS,5           ;Bank1
                MOVLW       B'11111111'
                MOVWF       TRISA              ;PortA is input

                MOVLW       B'00000000'
                MOVWF       TRISB              ;PortB is output

                MOVLW       B'11111111'
                MOVWF       TRISC              ;PortC is input

                MOVLW       B'00000111'
                MOVWF       OPTION_R           ;Option Register, TMR0 / 256

                MOVLW       B'00000000'
                MOVWF       ADCON1             ;PortA bits 0, 1, 2, 3, 5 are analogue
                BSF         STATUS,6           ;BANK3
                BCF         EECON1,7           ;Data memory on.
                BCF         STATUS,5
                BCF         STATUS,6           ;BANK0 return
                BSF         ADCON0,0           ;turn on A/D.
                CLRF        PORTA
                CLRF        PORTB
                CLRF        PORTC
```

;***
;Program starts now.

;Check the temperature on AN0
```
BEGIN       BCF         ADCON0,CHS0        ;C to select AN0
                BSF         ADCON0,GODONE
WAIT1       BTFSC       ADCON0,GODONE
                GOTO        WAIT1
                MOVF        ADRES,W
                SUBLW       .163               ;163 – W
```

```
          BTFSS    STATUS,CARRY      ;C if W > 163 i.e. hot (above
                                      17°C)
                   GOTO     HEAT_OFF

          MOVF     ADRES,W
          SUBLW    .148              ;148 – W
          BTFSC    STATUS,CARRY      ;S if W < 148 i.e. cold
                                      (below 15°C)
                   GOTO     HEAT_ON

;Check the soil moisture on AN1
SOIL      BSF      ADCON0,CHS0       ;S to select AN1
          BSF      ADCON0,GODONE
WAIT2     BTFSC    ADCON0,GODONE
          GOTO     WAIT2
          MOVF     ADRES,W
          SUBLW    .133              ;133 – W
          BTFSS    STATUS,CARRY      ;C if W > 133 i.e. dry
          GOTO     WATER_ON


;Check water is above minimum
WATER     BTFSC    PORTC,0           ;C if below minimum
          GOTO     ALARM_OFF
          GOTO     ALARM_ON

END
```

Explanation of code

In the previous analogue circuits in Chapter 8 we only used one analogue input on AN0. We now have two analogue inputs on AN0 and AN1. When making an analogue measurement we must specify which analogue channel we wish to measure. The default is AN0 – when moving to AN1 we select AN1 by setting channel select bit 0, i.e. BSF ADCON0,CHS0; when moving back to AN0 clear the channel select bit. The eight channels, AN0 to AN7, are seclected using bits CHS2, CHS1, CHS0.

- The temperature is read on AN0 and then checked to see if it is greater than 17°C, by subtracting the A/D reading from 163 (the reading equating to 17°C). The Carry Bit in the status register indicates if the result is +ve or –ve being set or clear. We then goto turn off the heater if the temperature is above 17°C or check if the temperature is below 15°C. In which case we turn on the heater.

- The soil moisture is checked next. AN1 is selected and the reading compared, this time to 133 indicating dry soil. The program either goes to turn on the water valve if the soil is dry or continues to check the water level if the soil is wet.
- If the water level is below minimum then the alarm sounds, if above minimum the alarm is turned off. The program then repeats the checking of the inputs and reacts to them accordingly.

Programming the 16F872 microcontroller using PICSTART PLUS

Once the program GREENHO.ASM has been saved it is then assembled using MPASMWIN. The next step as previously is to program GREENHO.HEX into the micro using PICSTART PLUS.

This process has been outlined in Chapter 2, but there are a few more selections to attend to in the 'Device specification' section.

- Select the device 16F872, if this device is not available you will require a later version of MPLAB, obtainable from www.microchip.com.
- Set the fuses.

Configuration bits

The configuration bit settings when programming the 16F872 for the greenhouse program are shown in Figure 13.5.

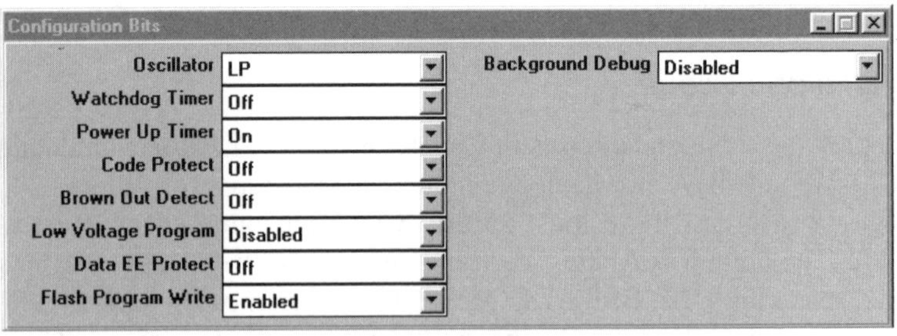

Figure 13.5 Greenhouse program configuration bits

Reconfiguring the 16F872 header

- The port settings are changed as they were for the 16F84, i.e. a 1 means the pin is an input and a 0 means an output.
- The Option register is configured as in the 16F84.
- The A/D converter configuration is adjusted using A/D configuration

register 1, i.e. ADCON1 shown in Figure 13.6.

ADFM			PCFG3	PCFG2	PCFG1	PCFG0
bit 7						bit 0

Figure 13.6 ADCON1, A/D Port Configuration Register 1

Bit 7 is the A/D Format Select bit, which selects which bits of the A/D result registers are used, i.e. the A/D can use 10 bits which require two result registers, ADRESH and ADRESL. Two formats are available: (a) the most significant bits of ADRESH read as 0, with ADFM = 1;

or (b) the least significant bits of ADRESL read as 0, with ADFM = 0

ADRESH

ADRESL

		0	0	0	0	0	0

For 8-bit operation condition (b) is used with ADRESH as the 8 most significant bits of the A/D result. This is the default configuration used in HEADER872.ASM where ADRESH (ADRES in the equates) is register 1Eh in Bank0.

Figure 13.7 shows the A/D Port Configuration settings for PCFG3, PCFG2, PCFG1 and PCFG0.

PCFG3:PCFG0	AN7 E2	AN6 E1	AN5 E0	AN4 A5	AN3 A3	AN2 A2	AN1 A1	AN0 A0	Vref+	Vref–
0000	A	A	A	A	A	A	A	A	Vdd	Vss
0001	A	A	A	A	Vref+	A	A	A	A3	Vss
0010	D	D	D	A	A	A	A	A	Vdd	Vss
0011	D	D	D	A	Vref+	A	A	A	A3	Vss
0100	D	D	D	D	A	D	A	A	Vdd	Vss
0101	D	D	D	D	Vref+	D	A	A	A3	Vss
011X	D	D	D	D	D	D	D	D	Vdd	Vss
1000	A	A	A	A	Vref+	Vref–	A	A	A3	A2
1001	D	D	A	A	A	A	A	A	Vdd	Vss
1010	D	D	A	A	Vref+	A	A	A	A3	Vss
1011	D	D	A	A	Vref+	Vref–	A	A	A3	A2
1100	D	D	D	A	Vref+	Vref–	A	A	A3	A2
1101	D	D	D	D	Vref+	Vref–	A	A	A3	A2
1110	D	D	D	D	D	D	D	A	Vdd	Vss
1111	D	D	D	D	Vref+	Vref–	D	A	A3	A2

A = analogue input, D = digital input
Vdd = +ve supply, Vss = –ve supply
Vref+ = high voltage reference
Vref– = low voltage reference
A3 = PortA,3 A2 = PortA,2

Note: AN7, AN6 and AN5 are only available on the 40 pin devices 16F871, 16F874 and 16F877.

Figure 13.7 A/D Port Configuration.

14
The 16F62X microcontroller

The 16F62X family of microcontrollers includes the two devices, 16F627 and 16F628. At the time of writing they are the latest editions to the PIC microcontroller range.

The 16F62X microcontrollers are flash devices and have 18 pins and data EEPROM just like the 16F84, but they have more functions. Notably there is an onboard oscillator so an external crystal is not required. This frees up two pins for extra I/O. The 16F62X in fact can use 16 of its 18 pins as I/O.

The 16F62X devices are set to replace the popular 16F84 device – not only do they have more functions, but they are also about 30% cheaper!

Figure 14.1 shows the specification of the 16F62X devices and the 16F84 for comparison.

Device	Flash Program Memory (bytes)	RAM Data Memory (bytes)	EEPROM Data Memory (bytes)	Timer Modules	I/O Pins
16F627	1024	224	128	3	16
16F628	2048	224	128	3	16
16F84	1024	68	64	1	13

Figure 14.1 The 16F62X specification

16F62X oscillator modes

The 16F62X can be operated in eight different oscillator modes. They are selected when programming the device just like the 16F84.

The options are:

- LP   Low Power Crystal, 32.768kHz
- XT   4MHz Crystal
- HS   High Speed Crystal, 20MHz
- ER   External Resistor (2 modes)

- INTRC Internal Resistor/Capacitor (2 modes)
- EC External Clock in

The two modes for the internal resistor/capacitor configuration are 4MHz and 37kHz. The default setting is 4MHz. The 16F627 header, HEAD62RC.ASM, selects the 37kHz oscillator by clearing the OSCF (oscillator frequency) bit, bit 3, in the Peripheral Control Register, PCON, with BCF PCON,3.

There was obviously a good reason for Microchip choosing 37kHz for the oscillator instead of 32.768kHz, I only wish I knew what it was! 32.768kHz as we have seen before (HEADER84.ASM), can give us TMR0 pulses of 32 a second when setting the OPTION register to divide the program timing pulses by 256.

With a 37kHz:

- Clock frequency = 37kHz.
- Program execution frequency is 37kHz / 4 = 9250Hz.
- Setting the prescaler to /32 gives TMR0 pulses of 9250 / 32 = 289.0625Hz = 0.03459459s for each pulse.
- Counting 29 TMR0 pulses gives a time of 0.100324324s, i.e. 0.1s + 0.3% error. If this error, about 4.5 minutes a day, is unacceptable then a 32.768kHz crystal can be used as we did with the 16F84.

Since the programs used previously on the 16F84 did not require any accurate timing our 16F62X header will set the prescaler to divide by 32 and use a subroutine to count 29 TMR0 pulses to give a time of 0.1s. All of the 16F84 programs can then be transferred to the 16F62X header. The choice of a 32.768kHz crystal or the 37kHz internal RC will obviously make a difference to the timing routines in the header. I have therefore included two headers for the 16F62X devices: HEAD62LP.ASM for use with the 32kHz crystal and HEAD62RC.ASM for use with the 37kHz internal RC oscillator.

16F62X and 16F84 pinouts

16F62X Pinout				16F84 Pinout			
A2	1	18	A1	A2	1	18	A1
A3	2	17	A0	A3	2	17	A0
A4/T0CLKIN	3	16	A7/OSC1/CLKIN	A4/T0CLKIN	3	16	OSC1/CLKIN
A5/MCLR	4	15	A6/OSC2/CLKOUT	MCLR	4	15	OSC2/CLKOUT
Vss	5	14	Vdd	Vss	5	14	Vdd
B0	6	13	B7	B0	6	13	B7
B1	7	12	B6	B1	7	12	B6
B2	8	11	B5	B2	8	11	B5
B3	9	10	B4	B3	9	10	B4

16F62X Port Configuration

The header HEAD62RC.ASM will configure the 16F62X I/O as shown in Figure 14.2.

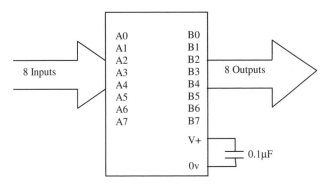

Figure 14.2 The 16F62X Port Configuration in HEAD62RC.ASM

The header (HEAD62LP.ASM) will configure the 16F62X I/O as shown in Figure 14.3

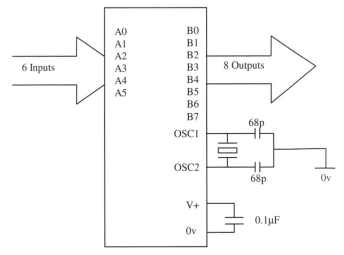

Figure 14.3 The 16F62X Port Configuration in HEAD62LP.ASM

16F62X memory map

The 16F62X memory map is shown in Figure 14.4.

Address	File Name	File Name	File Name	File Name
00h	Ind.Add	Ind.Add	Ind.Add	Ind.Add
01h	TMR0	Option	TMR0	Option
02h	PCL	PCL	PCL	PCL
03h	Status	Status	Status	Status
04h	FSR	FSR	FSR	FSR
05h	PORTA	TRISA		
06h	PORTB	TRISB	PORTB	TRISB
07h				
08h				
09h				
0Ah	PCLATH	PCLATH	PCLATH	PCLATH
0Bh	INTCON	INTCON	INTCON	INTCON
0Ch	PIR1	PIE1		
0Dh				
0Eh	TMR1L	PCON		
0Fh	TMR1H			
10h	T1CON			
11h	TMR2			
12h	T2CON	PR2		
13h				
14h				
15h	CCPR1L			
16h	CCPR1H			
17h	CCP1CON			
18h	RCSTA	TXSTA		
19h	TXREG	SPBRG		
1Ah	RCREG	EEDATA		
1Bh		EEADR		
1Ch		EECON1		
1Dh		EECON2		
1Eh				
1Fh	CMCON	VRCON		
.	General Purpose Register 96 bytes	General Purpose Register 80 bytes	General Purpose Register 48 bytes	
6Fh				
. 7Fh				
	Bank0	Bank1	Bank2	Bank3

Figure 14.4　16F62X memory map

The 16F62X headers
HEAD62LP.ASM

```
;HEAD62LP.ASM using the 32kHz crystal
                    ;PortA bits 0 to 5 are inputs
                    ;PortB bits 0 to 7 are outputs
                    ;Prescaler / 256
;*********************************************
;EQUATES SECTION

            TMR0        EQU     1
            OPTION_R    EQU     1
            PORTA       EQU     5
            PORTB       EQU     6
            TRISA       EQU     5
            TRISB       EQU     6
            STATUS      EQU     3
            ZEROBIT     EQU     2
            CARRY       EQU     0
            EEADR       EQU     1BH
            EEDATA      EQU     1AH
            EECON1      EQU     1CH
            EECON2      EQU     1DH
            RD          EQU     0
            WR          EQU     1
            WREN        EQU     2
            COUNT       EQU     20H
;*********************************************************
            LIST        P=16F627            ;using the 627
            ORG         0
            GOTO        START
;*********************************************************
;SUBROUTINE SECTION.

;1 SECOND DELAY
DELAY1      CLRF        TMR0                ;Start TMR0
LOOPA       MOVF        TMR0,W              ;Read TMR0 into W
            SUBLW       .32                 ;TIME - W
            BTFSS       STATUS,ZEROBIT      ;Check TIME-W=0
            GOTO        LOOPA
            RETLW       0                   ;Return after TMR0 = 32
```

```
;0.5 SECOND DELAY
DELAYP5    CLRF      TMR0              ;Start TMR0
LOOPB      MOVF      TMR0,W            ;Read TMR0 into W
           SUBLW     .16               ;TIME - W
           BTFSS     STATUS,ZEROBIT    ;Check TIME-W=0
           GOTO      LOOPB
           RETLW     0                 ;Return after TMR0 = 16
```

;**
;CONFIGURATION SECTION.

```
START      BSF       STATUS,5          ;Bank1
           MOVLW     B'11111111'
           MOVWF     TRISA             ;PortA is input

           MOVLW     B'00000000'
           MOVWF     TRISB             ;PortB is output

           MOVLW     B'00000111'
           MOVWF     OPTION_R          ;Option Register,
                                       ;TMR0 / 256

           BCF       STATUS,5          ;Bank0
           CLRF      PORTA
           CLRF      PORTB
```

;**
;Program starts now.

HEAD62RC.ASM

;HEAD62RC.ASM using the 37kHz internal RC
                    ;PortA bits 0 to 7 are inputs
                    ;PortB bits 0 to 7 are outputs
                    ;Prescaler / 32
;***
;EQUATES SECTION

```
           TMR0      EQU    1
           OPTION_R  EQU    1
           PORTA     EQU    5
           PORTB     EQU    6
           TRISA     EQU    5
           TRISB     EQU    6
```

```
                STATUS    EQU    3
                ZEROBIT   EQU    2
                CARRY     EQU    0
                EEADR     EQU    1BH
                EEDATA    EQU    1AH
                EECON1    EQU    1CH
                EECON2    EQU    1DH
                RD        EQU    0
                WR        EQU    1
                WREN      EQU    2
                PCON      EQU    0EH
                COUNT     EQU    20H
;*****************************************************
                LIST      P=16F627            ;using the 627
                ORG       0
                GOTO      START
;*********************************************************
;SUBROUTINE SECTION.


;0.1 SECOND DELAY
DELAYP1         CLRF      TMR0                ;Start TMR0
LOOPA           MOVF      TMR0,W              ;Read TMR0 into W
                SUBLW     .29                 ;TIME - W
                BTFSS     STATUS,ZEROBIT      ;Check TIME-W=0
                GOTO      LOOPA
                RETLW     0                   ;Return after TMR0 = 29


;0.5 SECOND DELAY
DELAYP5         MOVLW     5
                MOVWF     COUNT
LOOPB           CALL      DELAYP1             ;0.1s delay
                DECFSZ    COUNT
                GOTO      LOOPB
                RETLW     0                   ;Return after 5 DELAYP1


;1 SECOND DELAY
DELAY1          MOVLW     10
                MOVWF     COUNT
LOOPC           CALL      DELAYP1             ;0.1s delay
                DECFSZ    COUNT
                GOTO      LOOPC
                RETLW     0                   ;Return after 10 DELAYP1

;***********************************************************************
```

;CONFIGURATION SECTION.

```
START       BSF        STATUS,5          ;Bank1
            MOVLW      B'11111111'
            MOVWF      TRISA             ;PortA is input

            MOVLW      B'00000000'
            MOVWF      TRISB             ;PortB is output

            MOVLW      B'00000100'
            MOVWF      OPTION_R          ;Option Register, TMR0 / 32
            CLRF       PCON              ;Select 37kHz oscillator.
            BCF        STATUS,5          ;Bank0
            CLRF       PORTA
            CLRF       PORTB
```

;***
;Program starts now.

A 16F627 application – flashing an LED on and off

In order to introduce the operation of the 16F672 device we will consider the simple example of the single LED flashing on and off, which was introduced in Chapter 2.

The 16F627 will be operated in the INTRC mode using the internal 37kHz oscillator.

The circuit diagram for this is shown in Figure 14.5.

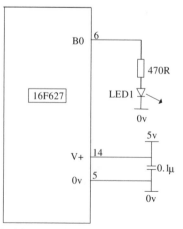

Figure 14.5 The 16F627 LED flashing circuit

The 16F627 LED flasher code

```
;FLASH_RC.ASM using the 37kHz internal RC
                    ;PortA bits 0 to 7 are inputs
                    ;PortB bits 0 to 7 are outputs
                    ;Prescaler / 32
;***********************************************
;EQUATES SECTION

            TMR0      EQU     1
            OPTION_R  EQU     1
            PORTA     EQU     5
            PORTB     EQU     6
            TRISA     EQU     5
            TRISB     EQU     6
            STATUS    EQU     3
            ZEROBIT   EQU     2
            CARRY     EQU     0
            EEADR     EQU     1BH
            EEDATA    EQU     1AH
            EECON1    EQU     1CH
            EECON2    EQU     1DH
            RD        EQU     0
            WR        EQU     1
            WREN      EQU     2
            PCON      EQU     0EH
            COUNT     EQU     20H
;*****************************************************************
            LIST      P=16F627            ;using the 627
            ORG       0
            GOTO      START
;*****************************************************************
;SUBROUTINE SECTION.

;0.1 SECOND DELAY
DELAYP1     CLRF      TMR0                ;Start TMR0
LOOPA       MOVF      TMR0,W              ;Read TMR0 into W
            SUBLW     .29                 ;TIME - W
            BTFSS     STATUS,ZEROBIT      ;Check TIME-W=0
            GOTO      LOOPA
            RETLW     0                   ;Return after TMR0 = 29
```

```
;0.5 SECOND DELAY
DELAYP5    MOVLW    5
           MOVWF    COUNT
LOOPB      CALL     DELAYP1          ;0.1s delay
           DECFSZ   COUNT
           GOTO     LOOPB
           RETLW    0                ;Return after 5 DELAYP1

;************************************************************
;CONFIGURATION SECTION.

START      BSF      STATUS,5         ;Bank1
           MOVLW    B'11111111'
           MOVWF    TRISA            ;PortA is input

           MOVLW    B'00000000'
           MOVWF    TRISB            ;PortB is output

           MOVLW    B'00000100'
           MOVWF    OPTION_R         ;Option Register, TMR0 / 32
           CLRF     PCON             ;Selects 37kHz oscillator.
           BCF      STATUS,5         ;Bank0
           CLRF     PORTA
           CLRF     PORTB

;************************************************************
;Program starts now.

BEGIN      BSF      PORTB,0          ;Turn on LED
           CALL     DELAYP5          ;Wait 0.5s
           BCF      PORTB,0          ;Turn off LED
           CALL     DELAYP5          ;Wait 0.5s
           GOTO     BEGIN            ;Repeat
END
```

The operation of the program after 'Program starts now' is exactly the same as in FLASHER.ASM in Chapter 2, using the 16F84.

All of the programs using the 16F84 can be transferred by copying the code starting at 'Program starts now' and pasting into HEAD62RC.ASM or HEAD62LP.ASM as required.

Configuration settings for the 16F627

When programming the code FLASH_RC.HEX into the 16F627 use the configuration settings shown in Figure 14.6.

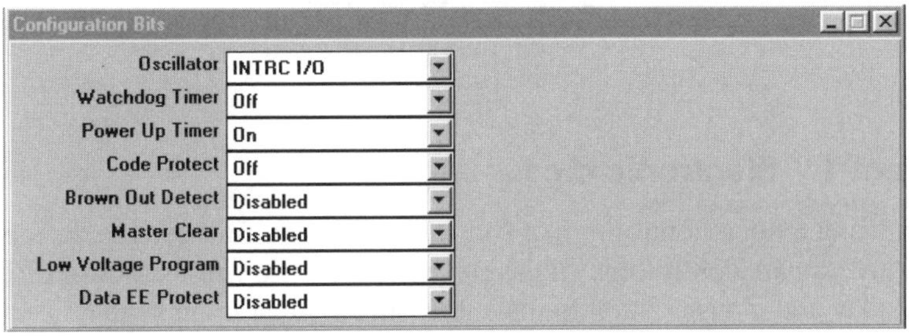

Figure 14.6 Configuration settings for FLASH_RC.HEX

Other features of the 16F62X

The 16F62X also includes:

- An analogue comparator module with two analogue comparators and an on-chip voltage reference module.
- Timer1, a 16-bit timer/counter module with external crystal/clock capability and Timer2, an 8-bit timer/counter with prescaler and postscaler.
- Capture, Compare and Pulse Width Modulation modes.

Please refer to the 16F62X data sheet for operation of these other features.

15
Projects

Project 1   Electronic dice

When using a microcontroller in a control system you first need to decide what hardware you are controlling. In the electronic dice we will use seven LEDs for the display and a push button to make the 'throw'. Just to make the dice a little more interesting we will use a buzzer to give an audible indication of the number thrown.

The circuit for the dice is shown in Figure 15.1. The push button is an input connected to PortA,2. The seven LEDs are connected to PortB and the buzzer is on A1.

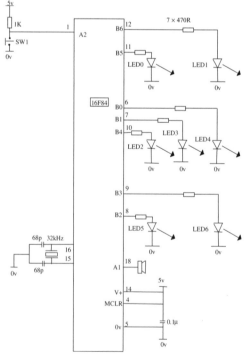

Figure 15.1 Circuit diagram for the electronic dice

The truth table for the dice is shown in Figure 15.2.

Throw	B7	B6	B5	B4	B3	B2	B1	B0
1	0	0	0	0	0	0	1	0
2	0	0	1	0	1	0	0	0
3	0	0	1	0	1	0	1	0
4	0	1	1	0	1	1	0	0
5	0	1	1	0	1	1	1	0
6	0	1	1	1	1	1	0	1

Figure 15.2 Truth table for the electronic dice

How does it work?

The dice has an input – the 'throw' button. When it is pressed the internal count repeatedly runs through from 1 to 6 changing some 8000 times a second and stops on a number when the button is released.

This would be a complicated circuit to design with timer, counter and decoder circuits. But now we can use one chip to do all the timing, counting and decoding functions. Not only that, I have also added a light flashing routine for the first few seconds when the dice is turned on. Try doing all that with one chip – other than a microcontroller.

The best way to describe the action of a program is with a flowchart. The flowchart for the dice is shown in Figure 15.3.

Program listing for the dice

The full program listing for the dice is given below in ;DICE.ASM.

;DICE.ASM

```
TMR0      EQU 1       ;TMR0 is FILE 1.
PC        EQU 2
PORTA     EQU 5       ;PORTA is FILE 5.
PORTB     EQU 6       ;PORTB is FILE 6.
STATUS    EQU 3       ;STATUS is FILE 3.
ZEROBIT   EQU 2       ;ZEROBIT is Bit 2.
COUNT     EQU 0CH     ;USER RAM LOCATION.
;****************************************************************
LIST      P=16F84     ;We are using the 16F84.
ORG       0           ;0 is the start address.
GOTO      START       ;goto start!

;****************************************************************
```

```
;SUBROUTINE SECTION.
;1 SECOND DELAY
DELAY1    CLRF     TMR0           ;Start TMR0
LOOPA     MOVF     TMR0,W         ;Read TMR0 in W
          SUBLW    .32            ;TIME - W
          BTFSS    STATUS,ZEROBIT ;Check TIME-W=0
          GOTO     LOOPA
          RETLW    0              ;Return after TMR0 = 32

;1/4 SECOND DELAY
DELAY     CLRF     TMR0           ;Start TMR0
LOOPB     MOVF     TMR0,W         ;Read TMR0 in W
          SUBLW    .8             ;TIME - W
          BTFSS    STATUS,ZEROBIT ;Check TIME-W=0
          GOTO     LOOPB
          RETLW    0              ;Return after TMR0 = 8

;P1 SECOND DELAY
DELAYP1   CLRF     TMR0           ;Start TMR0
LOOPC     MOVF     TMR0,W         ;Read TMR0 in W
          SUBLW    .3             ;TIME - W
          BTFSS    STATUS,ZEROBIT ;Check TIME-W=0
          GOTO     LOOPC
          RETLW    0              ;Return after TMR0 = 3
;************************************************************
;CONFIGURATION SECTION.

START     BSF      STATUS,5       ;Turn to BANK1
          MOVLW    B'00011101'    ;PORTA is mixed I/O.
          TRIS     PORTA
          MOVLW    0
          TRIS     PORTB          ;PORTB is Output
          MOVLW    B'00000111'
          OPTION                  ;PRESCALER is /256
          BCF      STATUS,5       ;Return to BANK0
          CLRF     PORTA          ;Clears PORTA
          CLRF     PORTB          ;Clears PORTB
;************************************************************
;Program starts now.

          CALL     DELAY1
          CALL     DELAY1
```

```
         CLRF     PORTB           ;Turn off LEDs and buzzer.

         MOVLW    .5
         MOVWF    COUNT

SEC1     MOVLW    60H             ;Light flashing routine.
         MOVWF    PORTB
         CALL     DELAY
         MOVLW    13H
         MOVWF    PORTB
         CALL     DELAY
         MOVLW    0CH
         MOVWF    PORTB
         CALL     DELAY
         MOVLW    13H
         MOVWF    PORTB
         CALL     DELAY
         DECFSZ   COUNT
         GOTO     SEC1

         CALL     DELAY1
         BSF      PORTA,1         ;Turn buzzer on
         CALL     DELAY1
         BCF      PORTA,1         ;Turn buzzer off

BEGIN    BTFSC    PORTA,2         ;Is switch pressed?
         GOTO     BEGIN           ;NO
         CALL     DELAYP1         ;YES
         CLRF     PORTB           ;Switch off LEDs
LOOP1    CLRF     TMR0            ;Start Timer
LOOP2    MOVF     TMR0,W          ;Put time into W.
         SUBLW    6               ;Is TMR0 = 6?
         BTFSC STATUS,ZEROBIT     ;Skip if TMR0 is not 6.
         GOTO     LOOP1           ;COUNT is 6, so reset timer.
         BTFSS    PORTA,2         ;skip if button released?
         GOTO     LOOP2           ;No, Carry on timing
         MOVF     COUNT,W         ;yes, put the COUNT into W.
         ADDWF    PC              ;Jump the value of W.
         GOTO     NUM1            ;COUNT=0
         GOTO     NUM2            ;COUNT=1
         GOTO     NUM3            ;COUNT=2
         GOTO     NUM4            ;COUNT=3
         GOTO     NUM5            ;COUNT=4
         GOTO     NUM6            ;COUNT=5
```

NUM1	MOVLW	B'00000010'	;Turn LED on
	MOVWF	PORTB	
	BSF	PORTA,1	;turn on buzzer for 1/4 sec.
	CALL	DELAY	
	BCF	PORTA,1	;Turn buzzer off.
	GOTO	BEGIN	;BEGIN AGAIN.
NUM2	MOVLW	B'00101000'	;TURN ON 2 LEDS.
	MOVWF	PORTB	
	BSF	PORTA,1	;turn on buzzer for 1/4 sec.
	CALL	DELAY	
	BCF	PORTA,1	;turn off buzzer for 1/4 sec.
	CALL	DELAY	
	BSF	PORTA,1	;turn on buzzer for 1/4 sec.
	CALL	DELAY	
	BCF	PORTA,1	Turn buzzer off.
	GOTO	BEGIN	
NUM3	MOVLW	B'00101010'	
	MOVWF	PORTB	
	BSF	PORTA,1	;turn on buzzer for 1/4 sec.
	CALL	DELAY	
	BCF	PORTA,1	;turn off buzzer for 1/4 sec.
	CALL	DELAY	
	BSF	PORTA,1	;turn on buzzer for 1/4 sec.
	CALL	DELAY	
	BCF	PORTA,1	;turn off buzzer for 1/4 sec.
	CALL	DELAY	
	BSF	PORTA,1	;turn on buzzer for 1/4 sec.
	CALL	DELAY	
	BCF	PORTA,1	;Turn off buzzer.
	GOTO	BEGIN	
NUM4	MOVLW	B'01101100'	
	MOVWF	PORTB	
	BSF	PORTA,1	;turn on buzzer for 1/4 sec.
	CALL	DELAY	
	BCF	PORTA,1	;turn off buzzer for 1/4 sec.
	CALL	DELAY	
	BSF	PORTA,1	;turn on buzzer for 1/4 sec.
	CALL	DELAY	
	BCF	PORTA,1	;turn off buzzer for 1/4 sec.
	CALL	DELAY	
	BSF	PORTA,1	;turn on buzzer for 1/4 sec.

```
         CALL      DELAY
         BCF       PORTA,1              ;turn off buzzer for 1/4 sec.
         CALL      DELAY
         BSF       PORTA,1              ;turn on buzzer for 1/4 sec.
         CALL      DELAY
         BCF       PORTA,1              ;Turn buzzer off.

         GOTO      BEGIN

NUM5     MOVLW     B'01101110'
         MOVWF     PORTB
         BSF       PORTA,1              ;turn on buzzer for 1/4 sec.
         CALL      DELAY
         BCF       PORTA,1              ;turn off buzzer for 1/4 sec.
         CALL      DELAY
         BSF       PORTA,1              ;turn on buzzer for 1/4 sec.
         CALL      DELAY
         BCF       PORTA,1              ;turn off buzzer for 1/4 sec.
         CALL      DELAY
         BSF       PORTA,1              ;turn on buzzer for 1/4 sec.
         CALL      DELAY
         BCF       PORTA,1              ;turn off buzzer for 1/4 sec.
         CALL      DELAY
         BSF       PORTA,1              ;turn on buzzer for 1/4 sec.
         CALL      DELAY
         BCF       PORTA,1              ;turn off buzzer for 1/4 sec.
         CALL      DELAY
         BSF       PORTA,1              ;turn on buzzer for 1/4 sec.
         CALL      DELAY
         BCF       PORTA,1              ;turn off buzzer.
         GOTO      BEGIN

NUM6     MOVLW     B'01111101'
         MOVWF     PORTB
         BSF       PORTA,1              ;turn on buzzer for 1/4 sec.
         CALL      DELAY
         BCF       PORTA,1              ;turn off buzzer for 1/4 sec.
         CALL      DELAY
         BSF       PORTA,1              ;turn on buzzer for 1/4 sec.
         CALL      DELAY
         BCF       PORTA,1              ;turn off buzzer for 1/4 sec.
         CALL      DELAY
         BSF       PORTA,1              ;turn on buzzer for 1/4 sec.
         CALL      DELAY
```

BCF	PORTA,1	;turn off buzzer for 1/4 sec.
CALL	DELAY	
BSF	PORTA,1	;turn on buzzer for 1/4 sec.
CALL	DELAY	
BCF	PORTA,1	;turn off buzzer for 1/4 sec.
CALL	DELAY	
BSF	PORTA,1	;turn on buzzer for 1/4 sec.
CALL	DELAY	
BCF	PORTA,1	;turn off buzzer for 1/4 sec.
CALL	DELAY	
BSF	PORTA,1	;turn on buzzer for 1/4 sec.
CALL	DELAY	
BCF	PORTA,1	;Turn buzzer off.
GOTO	BEGIN	

END

Modifications to the dice project

Can you think of any modifications you can make to this program? This would be a good way to learn how to use these chips. Perhaps you could add a roll routine so that a few numbers are shown before the dice finally comes to rest on the number. The initial display routine could also be customised. You could throw a 7.

Dice using 12C508

The dice circuit used eight outputs and one input – a total of nine I/O. But LEDs 0 and 6, 1 and 5, 2 and 4 work in pairs. I.e. they are on and off together. If these LEDs were paralleled up, then we only need six I/O, e.g.:

- Input from Switch
- Output to Buzzer
- Output to LEDs 0 and 6
- Output to LEDs 1 and 5
- Output to LEDs 2 and 4
- Output to LED 3

This project can then be undertaken using the six I/O of the 12C508.

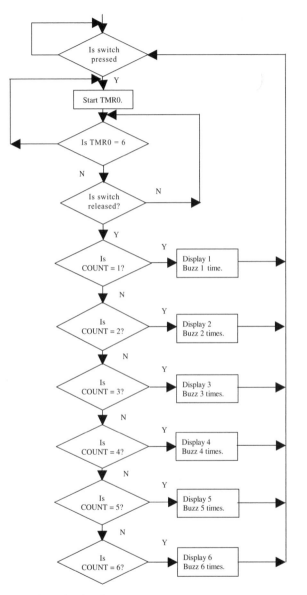

Figure 15.3 Flowchart for the dice

Project 2   Reaction timer

There are many question and answer games on the market that would benefit from a reaction timer which indicates the first player of a team to press. This project has the facility for up to six players. The circuit diagram for this project illustrated in Figure 15.4 uses six inputs and seven outputs.

Reaction timer operation

If B0 is the first to press B6 output LED lights
If B1 is the first to press B7 output LED lights
If B2 is the first to press A0 output LED lights
If B3 is the first to press A1 output LED lights
If B4 is the first to press A2 output LED lights
If B5 is the first to press A3 output LED lights
The buzzer is connected to A4.

The buzzer sounds for 4 seconds after a button is pressed. During this time no further presses are acknowledged. After the 4 seconds the buzzer stops and the LED is extinguished and the program resets.

The unit uses all 13 I/O but not all six button/LED combinations need be used. The program will not need altering.

Just one point in case you were wondering: B0–B5 have been used as inputs instead of PORTA because PORTB has internal pull-up resistors on the inputs. The switches do not need their own – no point in using five resistors if you don't have to.

The reaction timer program

;REACTION.ASM

```
TMR0        EQU       1          ;TMR0 is FILE 1.
PORTA       EQU       5          ;PORTA is FILE 5.
PORTB       EQU       6          ;PORTB is FILE 6.
STATUS      EQU       3          ;STATUS is FILE 3.
ZEROBIT     EQU       2          ;ZEROBIT is Bit 2.
COUNT       EQU       0CH        ;USER RAM LOCATION.
;*************************************************************
LIST        P=16F84              ;We are using the 16F84.
ORG         0                    ;0 is the start address.
GOTO        START                ;goto start!

;*************************************************************
```

;SUBROUTINE SECTION.

;1 SECOND DELAY
```
DELAY1      CLRF        TMR0                ;Start TMR0
LOOPA       MOVF        TMR0,W              ;Read TMR0 into W
            SUBLW       .32                 ;TIME - W
            BTFSS       STATUS,ZEROBIT      ;Check TIME-W=0
            GOTO        LOOPA
            RETLW       0                   ;Return after TMR0 = 32
```

;4 SECOND DELAY
```
DELAY4      CLRF        TMR0                ;Start TMR0
LOOPB       MOVF        TMR0,W              ;Read TMR0 into W
            SUBLW       .128                ;TIME - W
            BTFSS       STATUS,ZEROBIT      ;Check TIME-W=0
            GOTO        LOOPB
            RETLW       0                   ;Return after TMR0 = 128

ON0         BSF         PORTB,6             ;Turn on LED0
            BCF         PORTA,4             ;Turn on buzzer
            CALL        DELAY4              ;Wait 4 seconds
            BCF         PORTB,6             ;Turn off LED0
            BSF         PORTA,4             ;Turn off buzzer
            GOTO        SCAN

ON1         BSF         PORTB,7             ;Turn on LED1
            BCF         PORTA,4             ;Turn on buzzer
            CALL        DELAY4              ;Wait 4 seconds
            BCF         PORTB,7             ;Turn off LED1
            BSF         PORTA,4             ;Turn off buzzer
            GOTO        SCAN

ON2         BSF         PORTA,0             ;Turn on LED2
            BCF         PORTA,4             ;Turn on buzzer
            CALL        DELAY4              ;Wait 4 seconds
            BCF         PORTA,0             ;Turn off LED2
            BSF         PORTA,4             ;Turn off buzzer
            GOTO        SCAN

ON3         BSF         PORTA,1             ;Turn on LED3
            BCF         PORTA,4             ;Turn on buzzer
            CALL        DELAY4              ;Wait 4 seconds
            BCF         PORTA,1             ;Turn off LED3
            BSF         PORTA,4             ;Turn off buzzer
            GOTO        SCAN
```

```
ON4        BSF     PORTA,2          ;Turn on LED4
           BCF     PORTA,4          ;Turn on buzzer
           CALL    DELAY4           ;Wait 4 seconds
           BCF     PORTA,2          ;Turn off LED4
           BSF     PORTA,4          ;Turn off buzzer
           GOTO    SCAN

ON5        BSF     PORTA,3          ;Turn on LED5
           BCF     PORTA,4          ;Turn on buzzer
           CALL    DELAY4           ;Wait 4 seconds
           BCF     PORTA,3          ;Turn off LED5
           BSF     PORTA,4          ;Turn off buzzer
           GOTO    SCAN
;**********************************************************
;CONFIGURATION SECTION.
START      BSF     STATUS,5         ;Turn to BANK1
           MOVLW   B'00000000'      ;5 bits of PORTA are O/Ps.
           TRIS    PORTA
           MOVLW   B'00111111'
           TRIS    PORTB            ;PORTB IS MIXED I/O
           MOVLW   B'00000111'
           OPTION                   ;PRESCALER is /256
           BCF     STATUS,5         ;Return to BANK0
           CLRF    PORTA            ;Clears PORTA
           BSF     PORTA,4          ;Turn off buzzer (open drain)
           CLRF    PORTB            ;Clears PORTB
           CLRF    COUNT
;**********************************************************
;Program starts now.

           MOVLW   0FFH
           MOVWF   PORTA            ;Turn on PORTA outputs
           BCF     PORTA,4          ;Turn on buzzer (open drain)
           MOVWF   PORTB            ;Turn on PORTB outputs
           CALL    DELAY1           ;Wait 1 second
           CLRF    PORTA            ;Turn off PORTA outputs
           BSF     PORTA,4          ;Turn off buzzer (open drain)
           CLRF    PORTB            ;Turn off PORTB outputs

SCAN       BTFSS   PORTB,0          ;Has B0 been pressed
           GOTO    ON0              ;Yes
           BTFSS   PORTB,1          ;Has B1 been pressed
           GOTO    ON1              ;Yes
           BTFSS   PORTB,2          ;Has B2 been pressed
           GOTO    ON2              ;Yes
```

BTFSS	PORTB,3	;Has B3 been pressed
GOTO	ON3	;Yes
BTFSS	PORTB,4	;Has B4 been pressed
GOTO	ON4	;Yes
BTFSS	PORTB,5	;Has B5 been pressed
GOTO	ON5	;Yes
GOTO	SCAN	

END

How does it work?

Just before we start please note – PORTA,4 is an open drain output, this means it can only sink current, i.e. you turn the output on with a logic 0. The buzzer in this case must have its +ve connection connected to 5v. The program starts by turning all the LEDs and the buzzer on for 1 second to check they are all working.

The program then tests each input in turn starting with B0; if it is set, i.e. not pressed, the program skips and checks the next input. When the last input B5 is checked and it is not pressed then the program skips the next instruction and goes back to SCAN again.

If one of the inputs is pressed the program branches to the relevant subroutine to turn on the appropriate LED and buzzer for 4 seconds before returning to scan the switches again.

Reaction timer development

One way of making this program more interesting and to develop your programming skills – when a button is pressed have the outputs jump around B6, A0, A3, A1, A2 then B7 before landing on the correct output. You could also have a flashing light routine at the start of the program to check they are working; you could also pulse the buzzer. The buzzer could be made to beep a number of times to give an audible indication of who was first to press. Another modification you could make is – think of one yourself, I'm not doing all the work.

Project 3   Burglar alarm
Operation

The circuit for the burglar alarm is shown in Figure 15.5 using the 16F84. It uses two inputs, SW0 and SW1 which are both normally closed. They can represent door contacts, passive infrared sensor outputs, window contacts or tilt switches.

SW0 has a delay on it but SW1 is immediately active. Both switches can have additional switches wired in series with them to provide extra security cover. If SW1 is a window contact in a caravan it could have a tilt switch wired in series with it, so if the caravan was moved the siren would sound immediately.

SW0 and SW1 are connected to PORTB so pull-ups are not required.
A buzzer is used to indicate entry and exit delays on the alarm and a siren is connected to the micro via an IRF511 (power MOSFET).

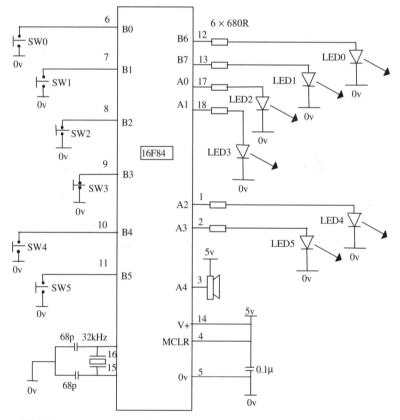

Figure 15.4 The reaction timer circuit

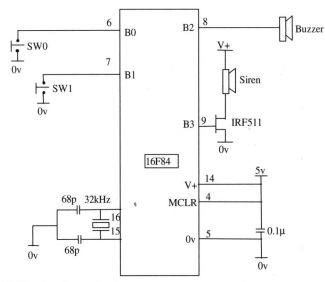

Figure 15.5 Burglar alarm circuit

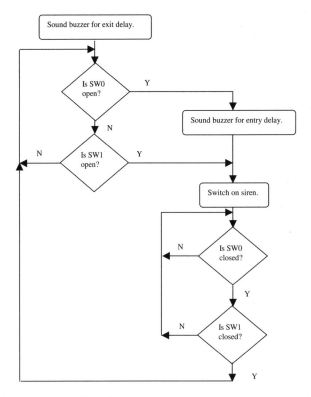

Figure 15.6 Burglar alarm flowchart

How does it work?

Consider the flow chart in Figure 15.6.

With reference to the flowchart: when the alarm is switched on a 30 second exit delay is activated and the buzzer sounds for this time. Switches 0 and 1 are continually checked until one of them is open.

If SW0 is opened a 30 second entry delay is activated and the buzzer sounds for this time, the siren will then sound for 5 minutes. If SW1 is opened the siren will sound immediately for 5 minutes.

The switches are then checked until they are both closed when the alarm resets to checking switches 0 and 1 until one of them opens again.

Switching off the power would disable the alarm.

Burglar alarm project code

The code for the burglar alarm is shown below in ALARM.ASM.

```
;ALARM.ASM

;EQUATES SECTION
TMR0        EQU        1          ;TMR0 is FILE 1.
PORTA       EQU        5          ;PORTA is FILE 5.
PORTB       EQU        6          ;PORTB is FILE 6.
STATUS      EQU        3          ;STATUS is FILE 3.
ZEROBIT     EQU        2          ;ZEROBIT is Bit 2.
COUNT       EQU        0CH        ;USER RAM LOCATION.
;*************************************************************
LIST        P=16F84    ;We are using the 16F84.
ORG         0          ;0 is the start address.
GOTO        START      ;goto start!


;*************************************************************
;SUBROUTINE SECTION.

;1 SECOND DELAY
DELAY1      CLRF       TMR0            ;Start TMR0
LOOPA       MOVF       TMR0,W          ;Read TMR0 into W
            SUBLW      .32             ;TIME - W
            BTFSS      STATUS,ZEROBIT  ;Check TIME-W=0
            GOTO       LOOPA
            RETLW      0               ;Return after TMR0 = 32
```

```
;0.5 SECOND DELAY
DELAYP5    CLRF      TMR0             ;Start TMR0
LOOPB      MOVF      TMR0,W           ;Read TMR0 into W
           SUBLW     .16              ;TIME - W
           BTFSS     STATUS,ZEROBIT   ;CHECK TIME-W=0
           GOTO      LOOPB
           RETLW     0                ;Return after TMR0 = 16

;0.25 SECOND DELAY
DELAYP25   CLRF      TMR0             ;Start TMR0
LOOPC      MOVF      TMR0,W           ;Read TMR0 into W
           SUBLW     .8               ;TIME - W
           BTFSS     STATUS,ZEROBIT   ;Check TIME-W=0
           GOTO      LOOPC
           RETLW     0                ;Return after TMR0 = 8

;5 SECOND DELAY
DELAY5     CLRF      TMR0             ;Start TMR0
LOOPD      MOVF      TMR0,W           ;Read TMR0 into W
           SUBLW     .160             ;TIME - W
           BTFSS     STATUS,ZEROBIT   ;Check TIME-W=0
           GOTO      LOOPD
           RETLW     0                ;Return after TMR0 = 160

BUZZER     MOVLW     .5
           MOVWF     COUNT            ;5 × 2 SECONDS
BUZZ1      BSF       PORTB,2
           CALL      DELAY1
           BCF       PORTB,2
           CALL      DELAY1
           DECFSZ    COUNT
           GOTO      BUZZ1
           MOVLW     .10
           MOVWF     COUNT            ;10 × 1 SECOND
BUZZ2      BSF       PORTB,2
           CALL      DELAYP5
           BCF       PORTB,2
           CALL      DELAYP5
           DECFSZ    COUNT
           GOTO      BUZZ2
           MOVLW     .20
           MOVWF     COUNT
BUZZ3      BSF       PORTB,2          ;20 × 0.5 SECONDS
```

```
              CALL       DELAYP25
              BCF        PORTB,2
              CALL       DELAYP25
              DECFSZ     COUNT
              GOTO       BUZZ3
              RETLW      0
```

;**
;CONFIGURATION SECTION.

```
START         BSF        STATUS,5        ;Turn to BANK1
              MOVLW      B'00011111'     ;5 bits of PORTA are I/Ps.
              TRIS       PORTA
              MOVLW      B'11110011'
              TRIS       PORTB           ;PORTB2,3 are OUTPUT
              MOVLW      B'00000111'
              OPTION                     ;PRESCALER is /256
              BCF        STATUS,5        ;Return to BANK0
              CLRF       PORTA           ;Clears PORTA
              CLRF       PORTB           ;Clears PORTB
              CLRF       COUNT
```

;**
;Program starts now.

```
              CALL       BUZZER          ;Exit delay
CHK_ON        BTFSC      PORTB,0         ;Check for alarm
              GOTO       ENTRY
              BTFSC      PORTB,1
              GOTO       SIREN
              GOTO       CHK_ON


ENTRY         CALL       BUZZER          ;Entry delay
SIREN         BSF        PORTB,3         ;5 minute siren
              MOVLW      .60
              MOVWF      COUNT
WAIT5         CALL       DELAY5
              DECFSZ     COUNT
              GOTO       WAIT5

              BCF        PORTB,3         ;Turn off Siren
CHK_OFF       BTFSC      PORTB,0         ;Check switches closed
              GOTO       CHK_OFF
              BTFSC      PORTB,1
              GOTO       CHK_OFF
```

```
              CALL        DELAYP25          ;antibounce
              GOTO        CHK_ON
END
```

The burglar alarm uses two inputs, and two outputs – a total of four I/O. We can therefore program the alarm with a 12C508 chip.

Burglar alarm using the 12C508

The circuit diagram for the alarm with the 12C508 is shown in Figure 15.7.

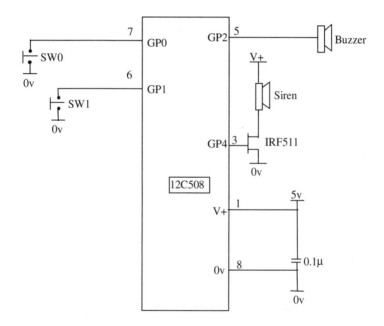

Figure 15.7 Burglar alarm using 12C508

Note in the circuit of Figure 15.7, showing the alarm using the 12C508, that no external oscillator circuit is required and that pull-ups are not required on pins GPIO,0 or GPIO,1 (or GPIO,3). Note: GPIO,3 is an input only pin. The flow-chart of course is the same. The code is shown below as ALARM_12.ASM using the header for the 12C508 from Chapter 12.

WARNING: The 12C508 only has a two level deep stack which means when you do a CALL you can only do one more CALL from that subroutine otherwise the program will get lost.

Program code for 12C508 burglar alarm

;ALARM_12.ASM FOR 12C508

```
TMRO        EQU        1          ;TMRO is FILE 1.
GPIO        EQU        6          ;GPIO is FILE 6.
OSCCAL      EQU        5          ;Oscillator calibration.
STATUS      EQU        3          ;STATUS is FILE 3.
ZEROBIT     EQU        2          ;ZEROBIT is Bit 2.
COUNT       EQU        07H        ;USER RAM LOCATION.
TIME        EQU        08H        ;TIME IS 39
COUNTB      EQU        09H
;********************************************************
LIST        P=12C508              ;We are using the 12C508.
            ORG        0          ;0 is the start address.
            GOTO       START      ;goto start!


;********************************************************

;SUBROUTINE SECTION.

;1 SECOND DELAY
DELAY1      MOVLW      .100       ;100 × 1/100 SEC.
            MOVWF      COUNT
TIMEA       CLRF       TMRO       ;Start TMRO
LOOPB       MOVF       TMRO,W     ;Read TMRO into W
            SUBWF      TIME,W     ;TIME - W
            BTFSS      STATUS,ZEROBIT    ;Check TIME-W=0
            GOTO       LOOPB
            DECFSZ     COUNT
            GOTO       TIMEA
            RETLW      0


;1/2 SECOND DELAY
DELAYP5     MOVLW      .50        ;50 × 1/100 SEC.
            MOVWF      COUNT
TIMEB       CLRF       TMRO       ;Start TMRO
LOOPC       MOVF       TMRO,W     ;Read TMRO into W
            SUBWF      TIME,W     ;TIME - W
            BTFSS      STATUS,ZEROBIT    ;CHECK TIME-W=0
            GOTO       LOOPC
            DECFSZ     COUNT
            GOTO       TIMEB
            RETLW      0
```

```
;1/4 SECOND DELAY
DELAYP25    MOVLW    .25              ;25 × 1/100 SEC.
            MOVWF    COUNT
TIMEC       CLRF     TMR0             ;Start TMR0
LOOPD       MOVF     TMR0,W           ;Read TMR0 IN W
            SUBWF    TIME,W           ;TIME - W
            BTFSS    STATUS,ZEROBIT   ;Check TIME-W=0
            GOTO     LOOPD
            DECFSZ   COUNT
            GOTO     TIMEC
            RETLW    0

;2 SECOND DELAY
DELAY2      MOVLW    .200             ;200 × 1/100 SEC.
            MOVWF    COUNT
TIMED       CLRF     TMR0             ;Start TMR0
LOOPE       MOVF     TMR0,W           ;Read TMR0 IN W
            SUBWF    TIME,W           ;TIME - W
            BTFSS    STATUS,ZEROBIT   ;Check TIME-W=0
            GOTO     LOOPE
            DECFSZ   COUNT
            GOTO     TIMED
            RETLW    0

BUZZER      MOVLW    .5
            MOVWF    COUNTB           ;5 × 2 Seconds
BUZZ1       BSF      GPIO,2
            CALL     DELAY1
            BCF      GPIO,2
            CALL     DELAY1
            DECFSZ   COUNTB
            GOTO     BUZZ1
            MOVLW    .10
            MOVWF    COUNTB           ;10 × 1 Second
BUZZ2       BSF      GPIO,2
            CALL     DELAYP5
            BCF      GPIO,2
            CALL     DELAYP5
            DECFSZ   COUNTB
            GOTO     BUZZ2
            MOVLW    .20
            MOVWF    COUNTB
BUZZ3       BSF      GPIO,2           ;20 × 0.5 Seconds
            CALL     DELAYP25
```

```
            BCF         GPIO,2
            CALL        DELAYP25
            DECFSZ      COUNTB
            GOTO        BUZZ3
            RETLW       0
;**********************************************************
;CONFIGURATION SECTION.

START       MOVWF       OSCCAL
            MOVLW       B'00101011'         ;GPIO bits 2 and 4 are O/Ps.
            TRIS        GPIO
            MOVLW       B'00000111'
            OPTION                          ;PRESCALER is /256
            CLRF        GPIO                ;Clears GPIO
            MOVLW       .39
            MOVWF       TIME


;**********************************************************
;Program starts now.


            CALL        BUZZER              ;Exit delay
CHK_ON      BTFSC       GPIO,0              ;Check for alarm
            GOTO        ENTRY
            BTFSC       GPIO,1
            GOTO        SIREN
            GOTO        CHK_ON


ENTRY       CALL        BUZZER              ;Entry delay
SIREN       BSF         GPIO,4              ;5 minute siren
            MOVLW       .150
            MOVWF       COUNTB
WAIT5       CALL        DELAY               ;150 × 2 seconds
            DECFSZ      COUNTB
            GOTO        WAIT5
            BCF         GPIO,4              ;Turn siren off


CHK_OFF     BTFSC       GPIO,0              ;Check switches closed
            GOTO        CHK_OFF
            BTFSC       GPIO,1
            GOTO        CHK_OFF
            CALL        DELAYP25            ;antibounce
            GOTO        CHK_ON
END
```

Fault finding

What if it all goes wrong!

The block diagram of the microcontroller in Figure 15.8 shows three sections: inputs, the microcontroller and outputs.

The microcontroller makes the output respond to changes in the inputs under program control.

Figure 15.8 Block diagram of the microcontroller circuit

All microcontroller circuits will have outputs and most will have inputs.

Checking inputs

If the inputs are not providing the correct signals to the micro then the outputs will not respond correctly.

Before checking inputs or outputs it is best to remove the microcontroller from the circuit – with the power switched off. Insert the micro in an IC holder so that it can be removed easily! This is essential for development work.

In order to check the inputs and outputs to the microcontroller let us consider a circuit we have looked at before in Chapter 3, the switch scanning circuit, shown below in Figure 15.9.

The four switches SW0, SW1, SW2 and SW3 turned on LED0, LED1, LED2 and LED3 respectively. To test the inputs monitor the voltage on the input pins to the microcontroller, pins 1, 2, 17 and 18. They should go high and low as you throw the switches.

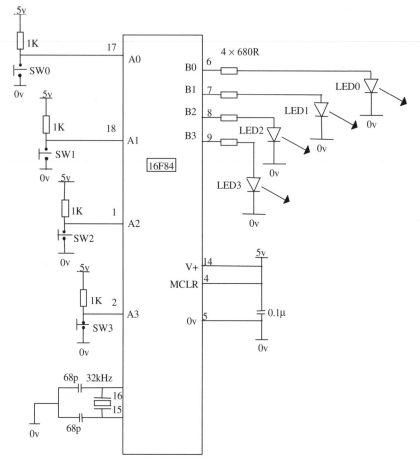

Figure 15.9 The switch scanning circuit

Checking outputs

The microcontroller will output 5v to turn on the outputs. To make sure the outputs are connected correctly, apply 5v to each output pin in turn to make sure the corresponding LED lights.

When 5v is applied to pin 6, the B0 output, then LED0 should light, etc. If it doesn't the resistor value could be incorrect or the LED faulty or inserted the wrong way round.

Check the supply voltage

Once you are sure the inputs and outputs are OK, check that the correct voltages are going to the pins. 5v on Vdd, pin 14 and MCLR, pin 4 and 0v on Vss, pin 5.

Check the oscillator

Check the oscillator is operating by monitoring the signal on CLKOUT, pin 15, with an oscilloscope or counter. Correct selection of the oscillator capacitor values are important – use 68pF with the 16C54 and 16F84 when using a 32kHz crystal.

Has the micro been programmed for the correct oscillator: R-C, LP, XT or HS? Most programs in this book use the LP configuration for the 32kHz Oscillator.

If everything is OK so far then the fault is with the microcontroller chip or the program.

Checking the microcontroller

If the program is not running it could be that you have a faulty microcontroller. You could of course try another, but how do you know that is a good one. The best course of action is to load a program you know works into the micro, such as FLASHER.ASM from Chapter 2. This flashes an LED on and off for 1 second; it doesn't use any inputs and only one output B0.

Checking the code

If there are no hardware faults then the problem is in your code. I find a useful aid is first to turn an LED on for 1 second and then turn it off. When this works you know that the microcontroller is OK, and that your timing has been set correctly and the oscillator and power supply are functioning correctly. With the switch scanning circuit you could turn all four LEDs on for 1 second anyway to serve as an LED check. To check your code, break it up into sections. Look at where the program stops running to identify the problem area.

If possible turn on LEDs on the outputs to indicate where you are in the program. If you are supposed to turn LED3 on when you go into a certain section of code and LED3 doesn't turn on, then of course you have not gone into that section and you are stuck somewhere else. These instructions can be removed later when the program is working.

Using a simulator

By using a simulator such as the Arizona MPSIM you can single step through the program and check it out a line at a time. MPSIM is part of MPLAB that is provided free from Microchip, the PIC manufacturers. MPSIM comes with a tutorial and can be downloaded from Microchip's website at www.microchip.com.

Common faults

Here are just a few daft things my students (or I!) have done:

- Not switched the power on.
- Put the chip in upside down.
- Programmed the wrong program into the micro.
- Corrected faults in the code but forgot to assemble it again, thus blowing the previous incorrect hex file again.
- Programmed incorrect fuses, i.e. watchdog timer and oscillator.

MMU PIC microcontroller development kit

There are a number of development kits on the market (and you can make your own). They have a socket for your micro, inputs and outputs that you can connect to your micro. They are ideal for program development. Once verified using the kit, if the system does not work then your circuit is at fault. I have developed such a kit at MMU shown in Figure 15.10 – details are also on the Manchester Metropolitan University website at:

http://www.mmu.ac.uk/c-a/edu/dandthome/picmicro/welcome.htm.

Figure 15.10 MMU PIC microcontroller development kit

16
Instruction set, files and registers

Microcontrollers work essentially by manipulating data in memory locations. Some of these memory locations are special registers, others are user files. In a control application data may be read from an input port, manipulated and passed to an output port.

To use the microcontroller you need to know how to move and manipulate this data in the memory. There are 35 instructions in the PIC 16F84 to enable you to do this. Using the microcontroller, then, is about using these instructions in a program. Like any vocabulary you do not use all the words all of the time; some you never use others only now and again. The PIC instruction set is like this – you can probably manage quite well with, say, 15 instructions.

Most of these instructions involve the use of the Working register or W reg. The W register is at the heart of the PIC microcontroller. To move data from file A to file B you have to move it from file A to W and then from W to file B, rather like a telephone system routes one caller to another via the exchange. The W reg also does the arithmetic and logical manipulating on the data.

The PIC microcontroller instruction set

To communicate with the PIC microcontroller you have to learn how to program it using its instruction set. The 16F84 chip has a 1k × 14 bit word EEP-ROM program memory, 68 × 8-bit general purpose registers and a 35 word instruction set made up of three groups of instructions, bit, byte and literal and control operations.

The instructions can be subdivided into three types:

- Bit instructions, which act on 1 bit in a file.
- Byte instructions, which act on all 8 bits in a file.
- Literal and control operations, which modify files with variables or control the movement of data from one file to another.

Bit instructions

The bit instructions act on a particular bit in a file, so the instruction would be followed by the data that specifies the file number and bit number, i.e. BSF 6,3. This code is not too informative so we would use something like BSF PORTB,BUZZER where PORTB is file 6 and the buzzer is connected to bit 3 of the output port. In the Equates Section we would see PORTB EQU 6 and BUZZER EQU 3.

BCF	Bit Clear in File.
BSF	Bit Set in File.
BTFSC	Bit Test in File Skip if Clear.
BTFSS	Bit Test in File Skip if Set.

Byte instructions

Byte instructions work on all 8 bits in the file. So a byte instruction would be followed by the appropriate file number, i.e. DECF 0CH. This statement is not too informative so we would again indicate the name of the file such as DECF COUNT. Of course we would need to declare in the Equates Section that COUNT was file 0CH, by COUNT EQU 0CH.

ADDWF	ADD W and F.
ANDWF	AND W and F.
CLRF	CLeaR File.
CLRW	CLeaR Working register.
COMF	COMplement File.
DECF	DECrement File.
DECFSZ	DECrement File Skip if Zero.
INCF	INCrement File.
INCFSZ	INCrement File Skip if Zero.
IORWF	Inclusive-OR W and F.
MOVF	MOVe F to W.
MOVWF	MOVe W to F.
NOP	No OPeration.
RLF	Rotate File one place Left.
RRF	Rotate File one place Right.
SUBWF	SUBtract W from F.
SWAPF	SWAp halves of F.
XORWF	eXclusive-OR W and F.

Literal and control operations

Literal and control operations manipulate data and perform program branching (jumps).

ADDLW	ADD Literal with W.
ANDLW	AND Literal with W.
CALL	CALL subroutine.
CLRWDT	CLeaR watchdog Timer.
GOTO	GOTO address.
IORLW	Inclusive-OR Literal with W.
MOVLW	MOVe Literal to W.
RETFIE	RETurn From IntErrupt.
RETLW	RETurn place Literal in W.
RETURN	RETURN from subroutine.
SLEEP	Go into standby mode.
SUBLW	SUBtract Literal from W.
XORLW	eXclusive-OR Literal and W.

These instructions operate mainly on two 8-bit registers – the Working register or W register and a file F which can be one of the 15 special registers or one of the 68 general purpose file registers that form the user memory (RAM) of the 16F84.

The memory map of the 16F84 is shown in Figure 16.1.

16F84 memory map

FILE ADDRESS	FILE NAME	FILE NAME
00	INDIRECT ADDRESS	INDIRECT ADDRESS
01	TMR0	OPTION
02	PCL	PCL
03	STATUS	STATUS
04	FSR	FSR
05	PORTA	TRISA
06	PORTB	TRISB
07	-	-
08	EEDATA	EECON1
09	EEADR	EECON2
0A	PCLATH	PCLATH
0B	INTCON	INTCON
0C	68	
	USER	
4F	FILES	
	BANK0	BANK1

Figure 16.1 16F84 memory map

The PIC microcontrollers are 8-bit devices – this means that the maximum number that can be stored in any one memory location is 255. Some PICs such as the 17C43 have 454 bytes of data memory. So to address memory locations greater than 255 the idea of pages or Banks has been introduced. Bank0 holds address locations up to 255, while Bank1 can hold a further 255, and Bank2 a further 255, etc. So you need to know what Bank a particular register or file is in. Banks are not used in the 16C54.

Registers

Registers are made up of 8 bits as shown in Figure 16.2.

bit 7	bit 6	bit 5	bit 4	bit 3	bit 2	bit 1	bit 0
1	0	1	1	0	0	1	0

MSB ---LSB

Figure 16.2 Register layout

Bit 0 is the Least Significant Bit (LSB) and bit 7 is the Most Significant Bit (MSB).

Register 00 indirect data addressing register

See Register 04 FSR, file select register.

Register 01 TMR0, TIMER 0/counter register

This register can be written to or read like any other register. It is used for counting or timing events. The contents of the register can be incremented (add 1) by the application of an external pulse applied to the TOCKI pin, i.e. counting cars into a car park or by the internal instruction cycle clock which runs at ¼ of the crystal frequency to time events.

Register 02 PCL, program counter

The program counter automatically increments to execute program instructions. An application of the use of the program counter is illustrated in the section on the look up table, Chapter 5 Program Examples.

Register 03, status register

The status register contains the result of the arithmetic or logical operations of the program. The 8 bits of the status register are shown in Figure 16.3.

bit 7	bit 6	bit 5	bit 4	bit 3	bit 2	bit 1	bit 0
IRP	RP1	RP0	TO	PD	Z	DC	C

Figure 16.3 Status register

- Bit 0, C, Carry Bit. This is set to a 1 if there is a carry from an addition or subtraction instruction, e.g. if one 8-bit number is added to another:

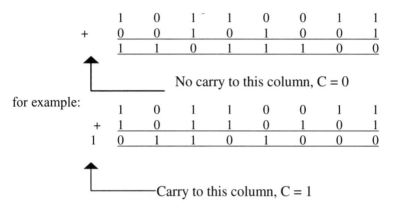

No carry to this column, C = 0

for example:

Carry to this column, C = 1

If the result of a subtraction is +ve or zero then the carry bit is set.
If the result of a subtraction is –ve then the carry bit is clear.

- Bit 2, Z, zerobit. This is set if the result of an arithmetic or logic operation is zero, i.e. countdown to zero. An important use of this bit is checking if a variable in memory is equal to a fixed value, i.e. does file CARS contain 150?

```
MOVLW    .150              ;Put 150 in W
SUBWF    CARS,W            ;Subtract W from CARS, i.e.
                           ;CARS-150
BTFSS    STATUS,ZEROBIT    ;Zerobit set if CARS=150
```

Register 04 FSR, file select register

The file select register is used in conjunction with the indirect data addressing register, register 00. They are used in indirect addressing to read or write data not from a specific file, but to or from a file indicated by the data in the file select register.

Register 05 PORT A and register 06 PORT B

Ports are the pin connections that allow the microcontroller to communicate with its surroundings. Port A is a 5-bit port on the 16F84, only the five LSBs are used. Port A bit 0 can also be programmed to be a clock input (T0CKI). Port B is an 8-bit port. To set up a port the instruction TRIS is used.

W register

The W register holds the result of an operation or an internal data transfer. It is like a telephone exchange – data comes into the W register and is transferred to another file.

OPTION register

This register is used to prescale the real time clock/counter. TMR0 clock runs at ¼ of the crystal frequency but can be divided by the prescaler for longer time measurements.

Stack

Stack is the name given to the memory location that keeps track of the program address when a CALL instruction is made. There is an eight level stack in the 16F84, which means that the program can jump to a subroutine and from there jump to another subroutine, making eight jumps in total, and the stack will be able to return it to the program. The 16C54 has a two level stack.

Instruction set summary

ADDLW   Adds a number to W, e.g. ADDLW 7 will add 7 to W, the result is placed in W.

ADDWF   Adds the contents of W to F, e.g. ADDWF 7 will add the contents of the W register and file 7. Note: the result is placed in file 7, e.g. ADDWF 7,W. The result is placed in W.
Status affected C, DC, and Z.

ANDLW   The contents of W are ANDed with an 8-bit number (literal). The result is placed in W, e.g. ANDLW 12H or ANDLW B'00010010' or ANDLW .18.
Status affected Z.

ANDWF   The contents of W are ANDed with F, e.g. ANDWF 12,W. The contents of file 12 are ANDed to the contents of W. Note: The result is placed in W, e.g. ANDWF 12. The result is placed in file 12.
Status affected Z.

BCF	Clear the bit in file F, e.g. BCF 6,4 bit 4 is cleared in file 6. File 6 is port B, this clears bit 4, i.e. bit 4 = 0.
BSF	Set bit in file F, e.g. BSF 6,4, this sets bit 4 in file 6, i.e. bit 4 = 1.
BTFSC	Test bit in file skip if clear, e.g. BTFSC 3,2. This tests bit 2 in file 3 – if it is clear then the next instruction is missed. File 3 is the status register and bit 2 is the zero bit so the program jumps if the result of an instruction was not zero.
BTFSS	Test bit in file skip if set, e.g. BTFSS 3,2, if bit 2 in file 3 is set then the next instruction is skipped.
CALL	This calls a subroutine in a program, e.g. CALL WAIT1MIN. This will call a routine (you have written) to wait for 1 minute. Maybe to turn a lamp on for 1 minute, and then return to the program.
CLRF	This clears file F; i.e. all 8 bits in file F are cleared, e.g. CLRF 5. Status affected Z.
CLRW	This clears the W register. Status affected Z.
CLRWT	The watchdog timer is cleared. The watchdog is a safety device in the microcontroller if the program crashes. The watchdog timer times out then restarts the program. Status affected TO, PD.
COMF	The 8 bits in file F are complemented, i.e. inverted, e.g. COMF 6. Status affected Z.
DECF	Subtract 1 from file F. Useful for counting down to zero, e.g. DECF 12 will store the result in 12. DECF 12,W will store the result in W leaving 12 unchanged. Status affected Z.
DECFSZ	The contents of F are decremented and the next instruction is skipped if the result is zero, e.g. DECFZ 12.
GOTO	This is an unconditional jump to a specified location in the program, e.g. GOTO SIREN.

INCF   Add 1 to F. This value could then be compared to another to see
            if a total had been achieved, e.g. INCF 14.
            Status affected Z.

INCFSZ   Add 1 to F if the result is zero then skip the next instruction, e.g.
            INCFSZ 19.

IORLW   The contents of the W register are OR-ed with a literal, e.g.
            IORLW 27.

i.e. W =	1	0	0	1	1	0	1	1
L =	0	0	0	1	1	0	0	1
L+W =	1	0	0	1	1	0	1	1

            This is a very useful way of determining if any bit in a file is
            set, i.e. by OR-ing a file with 00000000. If all the bits in the file
            are zero the OR result is zero and the zerobit is set in the status
            register.
            Status affected Z.

IORWF   The contents of the W register are OR-ed with the file F, e.g.
            IORWF 7,W. The result is stored in W, e.g. IORWF 7. The result
            is stored in file 7.
            Status affected Z.

MOVF   The contents of the file F are moved into the W register, from
            there the data can be moved to an output port, e.g. MOVF 12,W.
            File 12 is moved to W, e.g. MOVF 12. File 12 is moved to file
            12? Zero is affected.
            Status affected Z.

MOVLW   The 8-bit literal is moved directly into W, e.g. MOVLW .127.
            Status affected Z.

MOVWF   The contents of the W register are moved to F, e.g. MOVWF 6.
            The data in the W register is placed on port B.

NOP   No operation – may seem like a daft idea but it is very useful for
            small delays. The NOP instruction delays for ¼ of the clock
            speed.

OPTION   The contents of W are loaded into the OPTION register. This
            instruction is used to prescale, i.e. set TMR0 timing rate as
            shown in Figure 16.4.

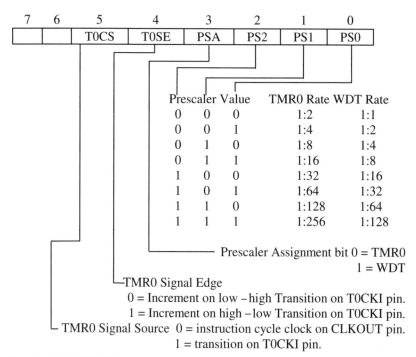

7	6	5	4	3	2	1	0
		T0CS	T0SE	PSA	PS2	PS1	PS0

Prescaler Value			TMR0 Rate	WDT Rate
0	0	0	1:2	1:1
0	0	1	1:4	1:2
0	1	0	1:8	1:4
0	1	1	1:16	1:8
1	0	0	1:32	1:16
1	0	1	1:64	1:32
1	1	0	1:128	1:64
1	1	1	1:256	1:128

Prescaler Assignment bit 0 = TMR0
1 = WDT

TMR0 Signal Edge
0 = Increment on low –high Transition on T0CKI pin.
1 = Increment on high –low Transition on T0CKI pin.
TMR0 Signal Source 0 = instruction cycle clock on CLKOUT pin.
1 = transition on T0CKI pin.

Figure 16.4 OPTION register

RETFIE   This instruction is used to return from an interrupt.

RETLW   This instruction is used at the end of a subroutine to return to the program following a CALL instruction. The literal value is placed in the W register. This instruction can also be used with a look up table, e.g. RETLW 0.

RETURN   This instruction is used to return from a subroutine.

RLF   The contents of the file F are rotated one place to the left through the carry flag. Shifting a binary number to the left means that the number has been multiplied by 2. This instruction is used when multiplying binary numbers, e.g. RLF 12,W. The result is placed in W, e.g. RLF 12. The result is placed in file 12. The diagram below shows file 12 being rotated left.

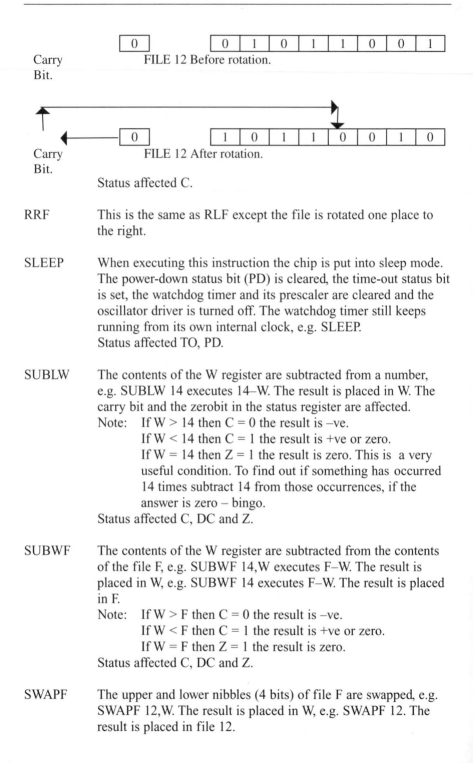

Carry Bit.   FILE 12 Before rotation.

Carry Bit.   FILE 12 After rotation.

Status affected C.

RRF   This is the same as RLF except the file is rotated one place to the right.

SLEEP   When executing this instruction the chip is put into sleep mode. The power-down status bit (PD) is cleared, the time-out status bit is set, the watchdog timer and its prescaler are cleared and the oscillator driver is turned off. The watchdog timer still keeps running from its own internal clock, e.g. SLEEP.
           Status affected TO, PD.

SUBLW   The contents of the W register are subtracted from a number, e.g. SUBLW 14 executes 14–W. The result is placed in W. The carry bit and the zerobit in the status register are affected.
           Note: If W > 14 then C = 0 the result is –ve.
                   If W < 14 then C = 1 the result is +ve or zero.
                   If W = 14 then Z = 1 the result is zero. This is a very useful condition. To find out if something has occurred 14 times subtract 14 from those occurrences, if the answer is zero – bingo.
           Status affected C, DC and Z.

SUBWF   The contents of the W register are subtracted from the contents of the file F, e.g. SUBWF 14,W executes F–W. The result is placed in W, e.g. SUBWF 14 executes F–W. The result is placed in F.
           Note: If W > F then C = 0 the result is –ve.
                   If W < F then C = 1 the result is +ve or zero.
                   If W = F then Z = 1 the result is zero.
           Status affected C, DC and Z.

SWAPF   The upper and lower nibbles (4 bits) of file F are swapped, e.g. SWAPF 12,W. The result is placed in W, e.g. SWAPF 12. The result is placed in file 12.

File 12 before SWAPF

0	1	1	0	1	1	0	1

File 12 after SWAPF

1	1	0	1	0	1	1	0

TRIS   Load the TRIS register. The contents of the W register are loaded into the TRIS register. This then configures an I/O port as input or output, e.g.

     MOVLW B'00001111'
     TRIS 6

This sets the four LSB's of port B as inputs and the four MSBs as outputs. Note: 1 for an input, 0 for an output.

XORLW   The contents of the W register are Exclusive OR-ed with the literal. If the result is zero then the contents match, i.e. if a number on the input port, indicating temperature, is the same as the literal then the result is zero and the zerobit is set, i.e. $0 \oplus 0 = 0$, $0 \oplus 1 = 1$, $1 \oplus 0 = 1$, $1 \oplus 1 = 0$, e.g. XORLW 67. Status affected Z.

XORWF   The contents of the W register are Exclusive OR-ed with the contents of the file F, i.e. if a number on the input port, indicating temperature, is the same as the W register then the result is zero and the zerobit is set. Note: You cannot Exclusive OR the input port directly with a file, you have to do this by loading the file into the W register with an MOVF instruction, e.g. XORWF 17,W. The result is placed in W, e.g. XORWF 17. The result is placed in 17. Status affected Z.

Did you notice how vital the W register is in the operation of the microcontroller?

Data cannot go directly from A to B, it goes from A to W and then from W to B.

Appendix A
Pinouts

16C54 Pinout

A2	1		18	A1
A3	2		17	A0
	3		16	OSC1/CLKIN
MCLR	4		15	OSC2/CLKOUT
Vss	5		14	Vdd
B0	6		13	B7
B1	7		12	B6
B2	8		11	B5
B3	9		10	B4

16F84 Pinout

A2	1		18	A1
A3	2		17	A0
A4/T0CLKIN	3		16	OSC1/CLKIN
MCLR	4		15	OSC2/CLKOUT
Vss	5		14	Vdd
B0	6		13	B7
B1	7		12	B6
B2	8		11	B5
B3	9		10	B4

16C71/711 Pinout

A2/AN2	1		18	A1/AN1
A3/AN3/VRef	2		17	A0/AN0
A4/T0CLKIN	3		16	OSC1/CLKIN
MCLR	4		15	OSC2/CLKOUT
Vss	5		14	Vdd
B0	6		13	B7
B1	7		12	B6
B2	8		11	B5
B3	9		10	B4

12C508/9 Pinout

Vdd	1		8	Vss
GP5/OSC1/CLKIN	2		7	Gp0
GP4/OSC2	3		6	Gp1
GP3/MCLR/Vpp	4		5	GP2/T0CKI

16F873/6 Pinout

MCLR/Vpp/THV	1		28	B7/PGD
A0/AN0	2		27	B6/PGC
A1/AN1	3		26	B5
A2/AN2/Vref-	4		25	B4
A3/AN3/Vref+	5		24	B3/PGM
A4/T0CKI	6		23	B2
A5/AN4/SS	7		22	B1
Vss	8		21	B0/INT
OSC1/CLKIN	9		20	Vdd
OSC2/CLKOUT	10		19	Vss
C0/T1OSO/T1CLKI	11		18	C7/RX/DT
C1/T1OSI/CCP2	12		17	C6/TX/CK
C2/CCP1	13		16	C5/SDO
C3/SCK/SCL	14		15	C4/SDI/SDA

16F874/7 Pinout

MCLR/Vpp/THV	1		40	B7/PGD
A0/AN0	2		39	B6/PGC
A1/AN1	3		38	B5
A2/AN2/Vref-	4		37	B4
A3/AN3/Vref+	5		36	B3/PGM
A4/T0CKI	6		35	B2
A5/AN4/SS	7		34	B1
E0/RD/AN5	8		33	B0/INT
E1/WR/AN6	9		32	Vdd
E2/CS/AN7	10		31	Vss
Vdd	11		30	D7/PSP7
Vss	12		29	D6/PSP6
OSC1/CLKIN	13		28	D5/PSP5
OSC2/CLKOUT	14		27	D4/PSP4
C0/T1OSO/T1CLKI	15		26	C7/RX/DT
C1/T1OSI/CCP2	16		25	C6/TX/CK
C2/CCP1	17		24	C5/SDO
C3/SCK/SCL	18		23	C4/SDI/SDA
D0/PSP0	19		22	D3/PSP3
D1/PSP1	20		21	D2/PSP2

Appendix B
Microcontroller data

Product	Program Memory		E²PROM Data Memory	RAM Bytes	8-Bit ADC Channels	I/O Ports	Timers	Max. Speed MHz
	Bytes	Words						
PIC12CXXX – 400ns Instruction Execution, 33/35 Instructions, 8 pin package, 4MHz Internal Oscillator, 4/5 Oscillator Selections								
PIC12C508	768	512x12	-	25	-	6	1-8 bit, 1-WDT	4
PIC12C509	1536	1024x12	-	41	-	6	1-8 bit, 1-WDT	4
PIC12CE518	768	512x12	16	25	-	6	1-8 bit, 1-WDT	4
PIC12CE519	1536	1024x12	16	41	-	6	1-8 bit, 1-WDT	4
PIC12C671	1792	1024x14	-	128	4	6	1-8 bit, 1-WDT	10
PIC12C672	3584	2048x14	-	128	4	6	1-8 bit, 1-WDT	10
PIC12CE673	1792	1024x14	16	128	4	6	1-8 bit, 1-WDT	10
PIC12CE674	3584	2048x14	16	128	4	6	1-8 bit, 1-WDT	10
PIC16C5X – 200ns Instruction Execution, 33 Instructions, 4/5 Oscillator Selections								
PIC16C52	576	384x12	-	25	-	12	1-8 bit, 1-WDT	4
PIC16C54	768	512x12	-	25	-	12	1-8 bit, 1-WDT	20
PIC16C55	768	512x12	-	24	-	20	1-8 bit, 1-WDT	20
PIC16C56	1536	1024x12	-	25	-	12	1-8 bit, 1-WDT	20
PIC16C57	3072	2048x12	-	72	-	20	1-8 bit, 1-WDT	20
PIC16C58A	3072	2048x12	-	73	-	12	1-8 bit, 1-WDT	20
PIC16C505	1536	1024x12	-	72	-	12	1-8 bit, 1-WDT	20

Product	Program Memory Bytes	Words	E²PROM Data Memory	RAM Bytes	8-Bit ADC Channels	I/O Ports	Timers	Max. Speed MHz
PIC16CXXX – 4-12 Interrupts, 200ns Instruction Execution, 35 Instructions, 4MHz Internal Oscillator, 4/5 Oscillator Selections								
PIC16C710	896	512×14	-	36	4	13	1-8 bit, 1-WDT	20
PIC16C71	1792	1024×14	-	36	4	13	1-8 bit, 1-WDT	20
PIC16C711	1792	1024×14	-	68	4	13	1-8 bit, 1-WDT	20
PIC16C712	1792	1024×14	-	128	4	13	1-16 bit, 2-8 bit, 1-WDT	20
PIC16C715	3584	2048×14	-	128	4	13	1-8 bit, 1-WDT	20
PIC16C716	3584	2048×14	-	128	4	13	1-16 bit, 2-8 bit, 1-WDT	20
PIC16C717	3584	2048×14	-	256	6 (10 bit)	16	1-16 bit, 2-8 bit, 1-WDT	20
PIC16C72	3584	2048×14	-	128	5	22	1-16 bit, 2-8 bit, 1-WDT	20
PIC16C73	7168	4096×14	-	192	5	22	1-16 bit, 2-8 bit, 1-WDT	20
PIC16C74	7168	4096×14	-	192	8	33	1-16 bit, 2-8 bit, 1-WDT	20
PIC16C76	14336	8192×14	-	368	5	22	1-16 bit, 2-8 bit, 1-WDT	20
PIC16C77	14336	8192×14	-	368	8	33	1-16 bit, 2-8 bit, 1-WDT	20
PIC16C770	3584	2048×14	-	256	6 (12 bit)	16	1-16 bit, 2-8 bit, 1-WDT	20
PIC16C771	7168	4096×14	-	256	6 (12 bit)	16	1-16 bit, 2-8 bit, 1-WDT	20
PIC16C773	7168	4096×14	-	256	6 (12 bit)	22	1-16 bit, 2-8 bit, 1-WDT	20
PIC16C774	7168	4096×14	-	256	10 (12 bit)	33	1-16 bit, 2-8 bit, 1-WDT	20
PIC16C745	14336	8192×14	-	256	5	22	1-16 bit, 2-8 bit, 1-WDT	24
PIC16C765	14336	8192×12	-	256	8	33	1-16 bit, 2-8 bit, 1-WDT	24

Product	Program Memory Bytes	Program Memory Words	E²PROM Data Memory	RAM Bytes	8-Bit ADC Channels	I/O Ports	Timers	Max. Speed MHz
PIC16CXXX – 4-12 Interrupts, 200ns Instruction Execution, 35 Instructions, 4MHz Internal Oscillator, 4/5 Oscillator Selection								
PIC1F83	896	512×14	64	36	-	13	1-8 bit, 1-WDT	10
PIC16F84	1792	1024×14	64	68	-	13	1-8 bit, 1-WDT	10
PIC16F872	3584	2048×14	64	128	5(10 bit)	22	1-16 bit, 2-8 bit, 1-WDT	20
PIC16F873	7168	4096×14	128	192	5(10 bit)	22	1-16 bit, 2-8 bit, 1-WDT	20
PIC16F874	7168	4096×14	128	192	8(10 bit)	33	1-16 bit, 2-8 bit, 1-WDT	20
PIC16F876	14336	8192×14	256	368	5(10 bit)	22	1-16 bit, 2-8 bit, 1-WDT	20
PIC16F877	14336	8192×14	256	368	8(10 bit)	33	1-16 bit, 2-8 bit, 1-WDT	20
PIC16C923	7168	4096×14	-	176	-	52	1-16 bit, 2-8 bit, 1-WDT	8
PIC16C924	7168	4096×14	-	176	5	52	1-16 bit, 2-8 bit, 1-WDT	8
PIC17CXXX – 4 12 Interrupts, 200ns Instruction Execution, 35 Instructions, 4MHz Internal Oscillator, 4/5 Oscillator Selection								
PIC17C42A	4096	2048×16	-	232	-	33	1-16 bit, 2-8 bit, 1-WDT	20
PIC17C43	8192	4096×16	-	454	-	33	1-16 bit, 2-8 bit, 1-WDT	20
PIC17C44	16384	8192×16	-	454	-	33	1-16 bit, 2-8 bit, 1-WDT	20
PIC17C752	16384	8192×16	-	256	6(12 bit)	16	1-16 bit, 2-8 bit, 1-WDT	20
PIC17C756	32768	16384×16	-	256	6(12 bit)	16	1-16 bit, 2-8 bit, 1-WDT	20
PIC17C762	16384	8192×16	-	256	6(12 bit)	22	1-16 bit, 2-8 bit, 1-WDT	20
PIC16C766	32768	16384×16	-	256	10(12 bit)	33	1-16 bit, 2-8 bit, 1-WDT	20
PIC18CXXX - 10 MIPS, 77 Instructions, C-compiler Efficient Instruction Set, Table Operation, Switchable Oscillator Sources								
PIC18C242	16384	8192×16	-	512	5(10 bit)	23	3-16 bit, 2-8 bit, 1-WDT	40
PIC18C442	16384	8192×16	-	512	8(10 bit)	34	3-16 bit, 2-8 bit, 1-WDT	40
PIC18C252	32768	16384×16	-	1536	5(10 bit)	23	3-16 bit, 2-8 bit, 1-WDT	40
PIC18C452	32768	16384×16	-	1536	8(10 bit)	34	3-16 bit, 2-8 bit, 1-WDT	40

Appendix C
Electrical characteristics

Absolute maximum ratings

Ambient temperature	−55°C to +125°C
Storage temperature	−65°C to +150°C
Voltage on any pin with respect to Vss (except Vdd and MCLR)	−0.6V to Vdd +0.6V
Voltage on Vdd with respect to Vss	0 to +7.5V
Voltage on MCLR with respect to Vss	0 to +14V
Total power dissipation	800mW
Max. current out of Vss pin	150mA
Max. current into Vdd pin (16C54)	50mA
Max. current into Vdd pin (16C71 and 16F84)	100mA
Max. current into an input pin	+/− 500mA
Max. output current sunk by any I/O pin	25mA
Max. output current sourced by any I/O pin	20mA
Max. output current sourced by a single I/O port (16C54)	40mA
Max. output current sourced by PORTA (16C71 and 16F84)	50mA
Max. output current sourced by PORTB (16C71 and 16F84)	100mA
Max. output current sunk by a single I/O port (16C54)	50mA
Max. output current sunk by PORTA (16C71 and 16F84)	80mA
Max. output current sunk by PORTB (16C71 and 16F84)	150mA

ts, numbers up to 65 535
ts, numbers up to 16 777 215
ts, numbers up to 4 294 967 295, etc.

tioned earlier hexadecimal numbers are a shorter way of writing binary
s. To do this divide the binary number into groups of four and write each
f four as a hex number:

10110 as 1001 0110 in binary
    =   9   6   in hex.

11010 as 1101 1010 in binary
    =   D   A   in hex.

D.2 shows some of the 255 numbers represented by 8 bits.

Decimal	Binary	Hexadecimal
0	00000000	00
1	00000001	01
2	00000010	02
3	0000011	03
4	00000100	04
5	00000101	05
8	00001000	08
15	00001111	0F
16	00010000	10
31	00011111	1F
32	00100000	20
50	00110010	32
63	00111111	3F
64	01000000	40
100	01100100	64
127	01111111	7F
128	10000000	80
150	10010110	96
200	11001000	C8
250	11111010	FA
251	11111011	FB
252	11111100	FC
253	11111101	FD
254	11111110	FE
255	11111111	FF

D.2 8-bit decimal, binary and hexadecimal representation

DC characteristics

Characteristic	Symbol	Min.	Typ.	Max.	Units	Conditions
Supply Voltage	Vdd					
PIC16C54-XT		3.0		6.25	V	Fosc = DC to 4MHz
PIC16C54-RC		3.0		6.25	V	Fosc = DC to 4MHz
PIC16C54-HS		4.5		5.5	V	Fosc = DC to 4MHz
PIC16C54-LP		2.5		6.25	V	Fosc = DC to 4MHz
RAM data retention voltage	Vdr	1.5			V	Device in Sleep Mode
Supply Current	Idd					
PIC16C54-XT			1.8	3.3	mA	Fosc = 4MHz, Vdd = 5.5V
PIC16C54-RC			1.8	3.3	mA	Fosc = 4MHz, Vdd = 5.5V
PIC16C54-HS			4.8	10	mA	Fosc = 10MHz, Vdd = 5.5V
			9.0	20	mA	Fosc = 20MHz, Vdd = 5.5V
PIC16C54-LP			15	32	µA	Fosc = 32kHz, Vdd =3.0V, WDT disabled.
Power down	Ipd		4	12	µA	Vdd = 3.0V, WDT enabled
Current			0.6	9	µA	Vdd = 3.0V, WDT disabled

Characteristic	Symbol	Min.	Typ.	Max.	Units	Conditions
Supply Voltage	Vdd					
PIC16C71-XT		4.0		6.0	V	Fosc = DC to 4MHz
PIC16C71-RC		4.0		6.0	V	Fosc = DC to 4MHz
PIC16C71-HS		4.5		5.5	V	Fosc = DC to 4MHz
PIC16C71-LP		4.0		6.0	V	Fosc = DC to 4MHz
RAM data retention voltage	Vdr	1.5			V	Device in Sleep Mode
Supply Current	Idd					
PIC16C71-XT			1.3	3.3	mA	Fosc = 4MHz, Vdd = 5.5V
PIC16C71-RC			1.3	3.3	mA	Fosc = 4MHz, Vdd = 5.5V
PIC16C71-HS			13.5	30	mA	Fosc = 20MHz, Vdd = 5.5V
PIC16C71-LP			35	70	µA	Fosc = 32kHz, Vdd = 3.0V, WDT disabled
Power down	Ipd		7	28	µA	Vdd = 4.0V, WDT enabled
Current			1	14	µA	Vdd = 4.0V, WDT disabled

Characteristic	Symbol	Min.	Typ.	Max.	Units	Conditions
Supply Voltage	Vdd					
PIC16F84-XT		4.0		6.0	V	Fosc = DC to 4MHz
PIC16F84-RC		4.0		6.0	V	Fosc = DC to 4MHz
PIC16F84-HS		4.5		5.5	V	Fosc = DC to 4MHz
PIC16F84-LP		4.0		6.0	V	Fosc = DC to 4MHz
RAM data retention voltage	Vdr	1.5			V	Device in Sleep Mode
Supply Current	Idd					
PIC16F84-XT			7.3	10	mA	Fosc = 4MHz, Vdd = 5.5V
PIC16F84-RC			7.3	10	mA	Fosc = 4MHz, Vdd = 5.5V
PIC16F84-HS			5	10	mA	Fosc = 10MHz, Vdd = 5.5V
PIC16F84-LP			35	400	µA	Fosc = 32kHz, Vdd =3.0V, WDT disabled
Power down	Ipd		40	100	µA	Vdd = 4.0V, WDT enabled
Current			38	100	µA	Vdd = 4.0V, WDT disabled

Appendix D
Decimal, binary and hexadecimal numbers

Homosapiens are used to decimal numbers, i.e. 0, 1, 2, 3 ... 9. Electronic machines or chips use Binary numbers 0 and 1 (OFF and ON).

Decimal numbers increase in tens, i.e. 267 means 7 ones, 6 tens and 2 hundreds.

100	10	1
2	6	7

Binary numbers increase in twos, i.e. 1010. The right hand 0 means no ones, the next digit means 1 two, the next means no fours, the next 1 eight, etc.

8	4	2	1
1	0	1	0

The binary number 1010 consists of 4 *binary digits* it is called a 4-bit number. 1010 is equivalent to 10 in decimal numbers.

We can change decimal numbers to binary and binary numbers to decimal. Digital systems, i.e. computers, are a little better than we are at this.

Consider the decimal number 89. To turn this into a binary number write the binary scale:

128	64	32	16	8	4	2	1

To make 89 we need $(0 \times 128) + (1 \times 64) + (0 \times 32) + (1 \times 16) + (1 \times 8) + (0 \times 4) + (1 \times 2) + (1 \times 1)$.

So 89 in decimal = 01011001 in binary.

To convert a binary number to decimal add up the various multiples of 2, i.e. 10011010 is:

128	64	32	16	8	4	2	1
1	0	0	1	1	0	1	0

$= 128 + 16 + 8 + 2 = 154$.

A long string of binary numbers is difficult to read, i.e. 11(shorter and therefore easier to put into a microcontroller h are used. Hexadecimal numbers increase in 16's and are de Figure D.1 shows these 16 digits and their decimal and bi

Decimal	Binary	Hexadecimal
0	0000	0
1	0001	1
2	0010	2
3	0011	3
4	0100	4
5	0101	5
6	0110	6
7	0111	7
8	1000	8
9	1001	9
10	1010	A
11	1011	B
12	1100	C
13	1101	D
14	1110	E
15	1111	F

Figure D.1 4 bit decimal, binary and hexadecimal representatio

The PIC microcontrollers are 8-bit micros, they use 8 bina representation such as 10010101. This is

128	64	32	16	8	4	2
1	0	0	1	0	1	0

$= 149$

The largest decimal number that can be represented by 11111111, which represents:-

128	64	32	16	8	4	2
1	1	1	1	1	1	1

$= 255$

But we can program our microcontroller to increase our nu from 8 bits, i.e. up to 255:

to 16 bi
to 24 bi
to 32 bi

As men
number
group o

i.e. 100

i.e. 110

Figure

Figure I

Appendix E
Useful contacts

- Author
d.w.smith@mmu.ac.uk

Manchester Metropolitan University, Design and Technology website
http://www.mmu.ac.uk/c-a/edu/dandthome/picmicro/welcome.htm

- A microcontroller design company
Electronic Controls Limited ☎ 01278 795 822
http://www.ecl-ltd.fsnet.co.uk

- Arizona Microchip, the company that manufactures the PICs. This website is a must:
http://www.MICROCHIP.COM

- Places to buy your components
Farnell ☎ 0113 263 6311 http://www.farnell.com
Rapid Electronics ☎ 01206 751166
RS Components ☎ 01536 444105 http://www.rs-components.com/rs
Maplin Electronics ☎ 01702 554000 http://www.maplin.co.uk

- A recommended magazine
Everyday Practical Electronics
http://www.epemag.wimborne.co.uk

Index